FISHTOWN

Leland, Michigan's Historic Fishery

To Bob-
Enjoy!
Laurie K Sommer
6-28-12

Arbutus Press

On the covers:

(FRONT) George Cook in Fishtown, with visitors watching, ca. 1950. COURTESY OF TRAVERSE AREA HISTORICAL SOCIETY

(BACK) Fishermen at the Fishtown net reel yard, ca. 1920. COURTESY OF LEELANAU HISTORICAL SOCIETY

Leland, Michigan's Historic Fishery

Laurie Kay Sommers

ARBUTUS PRESS
TRAVERSE CITY, MICHIGAN

Fishtown: Leland, Michigan's Historic Fishery

by Laurie Kay Sommers

Arbutus Press
Traverse City, Michigan
editor@arbutuspress.com
www.ArbutusPress.com

Book design by
Daniel Stewart / History by Design
dstewart@historybydesign.net

ISBN 978-1-933926-46-9

Library of Congress Cataloging-in-Publication Data available

Printed in the United States of America.

In loving memory of my father, who was the first Sommers to write about commercial fishing, and in honor of all the commercial fishermen of Fishtown.

Contents

Foreword

The Last Fishtown

by Amanda J. Holmes
Executive Director, Fishtown Preservation

I CAME TO KNOW FISHTOWN THROUGH SENSES AND STORIES, THE WAY MOST of us do. I've walked the docks several times a week year-round, through the flattening heat of summer, the piled snows of winter, and everything in between. I've also talked with hundreds of people, sometimes casually—often in Fishtown—and sometimes more formally, in interviews with a microphone in my hand. As I've come to know Fishtown, I've also gone searching the Great Lakes for our twin, for another survivor from the era when many small, wooden, weathered fishing villages hugged the lakeshores. For six years I've searched, and even as I've come to understand the singular role that Fishtown plays in the lives of so many—as one place with many meanings—I've also come to realize that there is no longer any other place like this. This place we love is the last Fishtown.

Our Fishtown, I'm happy to report, is very much alive. When I mention that I work for Fishtown Preservation, countless times the reply has come: "I *love* Fishtown!" This book was born out of that love. It is a celebration of this place and the people who made it, and are making it still. Fishtown has survived and grown through many kindnesses and much generosity, and in this book we hope to share with you some of what has been shared with us.

It's difficult—perhaps impossible—to capture in words the durable impression that Fishtown makes upon the senses. Tom Kelly, executive director of the Inland Seas Education Association and a man who has long lived around water, can't imagine Fishtown without the sound of the water cascading over the dam, the pulse beneath a river town. Mike Grosvenor, who for years operated Manitou Transit, the ferry service to the Manitou Islands, came to taste Fishtown as a boy through the saltiness of chub still hot from the smoker. Jeff Fisher, whose family has summered in Leland since 1903, also conjures Fishtown through the smell of smoked fish. "Maybe it's imprinted in our genes," he says, "but I think Fishtown draws people in." The sensory richness of the place—the way smells and sounds can bring back to life the moments we'd long forgotten—is probably why I see so many wedding parties in Fishtown, these wood walls photographed as the background for two clasped hands and the beginning of a shared future.

I've come to know Fishtown through the stories I've heard. I'm by training a folklorist, and shortly after the Fishtown Preservation Society (FPS) completed the purchase of Fishtown in 2007, I began interviewing fishermen and their families, local and summer residents, past shanty business owners, and many of the annual visitors bursting with their own Fishtown experiences. I've come to see Fishtown from these multiple perspectives and to understand that the story of Fishtown is one of collective knowledge and experiences both shared and personal.

This book began in those collected stories. In the months immediately following the purchase, FPS began a series of planning projects built upon the information

Fishtown fisherman Joel Petersen and Meggen Watt Petersen "take the plunge" in Fishtown, July 2011. PHOTO BY CORY WEBBER PHOTOGRAPHY, COURTESY OF JOEL AND MEGGEN WATT PETERSEN

gathered in these interviews. Project built upon project, culminating in the completion of an authoritative Historic Structure Report for Fishtown. This report, we realized, could be the basis for a much-needed history of Fishtown.

Laurie Kay Sommers, who headed the Historic Structure Report team, was the ideal person to write this book. Laurie is not only a folklorist like me, but also a musician, a tireless researcher and a storyteller. She holds a Ph.D. in Folklore, with a concentration in ethnomusicology, from Indiana University. Since 1981, she has worked as a public sector folklorist for organizations including the Indiana Division of State Parks, the Bureau of Florida Folklife, the Smithsonian Office of Folklife Programs, the Michigan State University Museum, and Valdosta State University College of the Arts, where she founded the South Georgia Folklife project. She is currently an independent consultant in folklore and historic preservation and a research associate for the Michigan State University Museum.

Laurie shared a fascination with the often dizzying details of how Fishtown has changed over the years. Once, in the middle of telling me about how she'd reconstructed the history of one of our peripatetic shanties, she said, "I can't think of anyone else who would care when that shanty was moved across the river!" I took this remark as Laurie meant it, as a great compliment. Laurie also savored her time with the fishermen, an occupation that either draws or creates great raconteurs, not only for their stories of past daily life in Fishtown, but also because she knows—as do they—that our links to the earlier days of Great Lakes commercial fishing grow more distant with each year.

Still, it's sometimes surprising where those links appear. Laurie and I have both enjoyed hearing stories about people who visited Fishtown as children and would bring their own children here, to pass on their affection into another generation. After she'd completed the Historic Structure Report, Laurie took some time to sort through the papers of her late father, Lawrence M. Sommers, who had been a professor of geography at Michigan State University. In her father's boxes Laurie discovered slides her father had taken during a visit to Fishtown in 1981. This confirmed what she had always sensed—that she was, while making her own story, also following in her father's footsteps.

We at the Fishtown Preservation Society hope that this book will increase your own appreciation for this place you may already love. We hope you will realize how rare this place is, and how Fishtown's survival is a story that continues with all of us. Fishtown has changed over the years—the buildings, the boats, and of course the faces. But there are constants, too, in buildings restored, in fish tugs repaired, in generations that return, and in the continuation of commercial fishing. That's why this book is a history, not an elegy.

The more I learn about Fishtown, the more I am devoted to its preservation. People who give to Fishtown do so because they get so much in return. For its sights, sounds, smells, and for the memories that live here, Fishtown is a place that matters. There is little more satisfying than being a part of securing its future.

Chapter 1

A Place Called Fishtown

"What a morning to spend in Fishtown! The air is filled with the aroma of smoked fish."
–*Leland Report*, 2001

On a blustery April morning, the door to Carlson's fishery in Leland's Fishtown opens with a blast of cold air. Two tourists, visiting Leelanau County, Michigan, on their spring break warm themselves from the chill as they take in their surroundings: the display case of fresh fish, employees processing the day's catch, the pungent smell of smoked fish. The walls of the sales area overflow with faded photos and newspaper clippings of the generations of Carlsons who have worked as commercial fishermen. Nestled within this gallery of Fishtown memories is a family treasure: a torn and yellowed commercial fishing license from 1909, owned by Swedish immigrant Nels Carlson, the first Carlson to ply the fishing grounds off Leland, and his Norwegian-born partner, Severt Johnson. Intrigued, the visitors ask, "What is this place?"

The answer is both simple and complex. Although known today as Fishtown, the fishing village on the Leland (historically Carp) River did not assume that name until the 1930s.[1] Fishtown today is a tourist destination, inspiration for artists, working waterfront, and centerpiece of a district listed on the National Register of Historic Places. In the shrinking world of Great Lakes commercial fisheries, Fishtown is a survivor, and Carlson's of Fishtown is its heartbeat. Of all the commercial fishing operations once based in Fishtown, only Carlson's, now

Alan Priest weighing freshly smoked fish for customers at Carlson's of Fishtown, 2010. PHOTO BY LAURIE KAY SOMMERS

solely a retail operation, remains. Since 2007, Carlson's and seven other shanties have been owned by the Fishtown Preservation Society (FPS), a not-for-profit organization dedicated to preserving the historical integrity of Fishtown as a living link to Great Lakes maritime heritage.

The "Historic Fishtown" sign atop the Ice House—along with the historic marker describing the Fishtown Historic District—declares that this is a place of history. But it is also one of the few remaining commercial fisheries on Lake Michigan. Above all, it is a place of stories. "The fishing is over for the season," quipped the *Leelanau Enterprise* newspaper in 1904, "but not the yarns." The stories are of place, tradition, and memory; of family and community in the face of adversity; of heroism, humor, luck, and tragedy on the Big Lake; of priceless fishing knowledge passed on from father to son; of artists and art students who captured the personalities, colors, and textures of Fishtown; of generations of summer visitors who infused the local economy with dollars even as the fishing ebbed and flowed. New stories emerge each day.

The Leland River flows through the heart of Fishtown and bears silent witness to these stories. Barely 4,000 feet in length, it connects Lake Leelanau (originally known as Carp Lake) to the east with Lake Michigan to the west. The portion that passes through Fishtown extends a scant 300 feet below the dam, which separates the upper river and Lake Leelanau from the outlet at Lake Michigan. River otters play in its clear green waters and, in the early autumn, large salmon rest on the sandy bottom, thwarted from swimming upstream to spawn by the six-foot waterfall over the dam. Pioneering settlers knew the river by the name "Carp River." Generations of fishermen called it "the fish creek."[2]

The river is Fishtown's soul. Without it, there would be no Fishtown. It is the reason early bands of Ottawa established a town at its mouth. Its water power led to Leland's founding by French Canadian millwright Antoine Manseau in 1853. Its waters sheltered fish tugs, pleasure craft, and the Manitou mail boat and ferry decades before Leland became a harbor of refuge. Children swam and played here, and generations of local residents and summer resorters fished from the

Fishtown during early spring, with the fish tug *Janice Sue* moored in the river downstream from the dam, 2011.
PHOTO BY LAURIE KAY SOMMERS

Jim VerSnyder whips up a batch of the popular whitefish pâté at Carlson's of Fishtown, 2010.
PHOTO BY LAURIE KAY SOMMERS

docks at water's edge. Along both sides of its banks stand the fish shanties that give Fishtown its name and identity.

The river, like Fishtown, has its seasons. On the April day when those tourists stumble upon Fishtown, the sky spits snow, and Lake Michigan is a wild froth of whitecaps. The frigid river ripples as the wind howls off the Big Lake. Fishtown's two remaining steel-hulled fish tugs, the trap-netter *Joy* and the gill-netter *Janice Sue*—the only craft in the river—rock at their moorings. They have just completed their first fishing runs of the year.

Six months later, on a warm September afternoon, the river at Fishtown will be full of boats. The commercial fishermen will have met their state-mandated quotas for whitefish and chubs, and the tugs will not go out again until early spring.[3] Most other boats at Fishtown will be fishing charters.

That September day is clear, and the islands dominate the western horizon, dense dark green forests broken by pale streaks of dune. Jim Munoz, captain of Leland's oldest charter service (established 1972), relaxes after a successful day with clients. He and another long-time charter captain, Jack Duffy, have just returned from fishing for trophy Chinook and Coho salmon and steelhead, lake, and brown trout off nearby North and South Manitou Islands. Lake Michigan has a light chop. In the distance, the *Manitou Isle* ferry motors across the Manitou Passage, the historic shipping lane that separates the islands and mainland. Currently owned by the fourth generation of the Grosvenor family of Leland, the ferry service is now called Manitou Island Transit. It operates under a license from the Sleeping Bear Dunes National Lakeshore and transports campers and day-trippers between Leland Harbor and the Manitous.

The *Manitou Isle* will dock in the harbor at the river's mouth, next to the Grosvenor's other boat, the newer and larger *Mishe-Mokwa*. Its passengers will join the tourists on Fishtown's north side, as they stroll the wooden docks, explore the retail shops that occupy former fish shanties, or enjoy a "Chubby Mary" at the Cove Restaurant's happy hour. The drink takes its name from the small Lake Michigan native fish, the chub, a local delicacy smoked by Carlson's of Fishtown. Chubs are rare these days—victims of the drastically changing lake biology—and sell out early. By late afternoon walk-in customers at Carlson's retail counter can only find a few whitefish and lake trout for sale, along with fish sausage, whitefish pâté, and smoked fish. Brine tanks contain the next day's batch of fish for smoking. Behind the retail cases, a crew surrounds the fillet table to dress a late-arriving load of whitefish trucked in from Petersen's

Fishtown, c. 2000. PHOTO BY KEITH BURNHAM

Fishery in Muskegon. The crew includes Nels Carlson—named for his great-great grandfather who was the first Carlson to fish commercially out of Leland; Nels is the fifth generation of Carlsons to work in the fishery. After the 1970s, Carlson's of Fishtown shifted more toward processing and retail, buying fish from other commercial fishermen to supplement its own catch and expanding its product line. When FPS purchased its Fishtown property in 2007, it also acquired the fish tugs and fishing licenses. The fishermen continue to process and sell the fish; FPS maintains the boats and hires the crew.

The Leland River divides the bustling north side of Fishtown, dominated by retail use, from the quiet south side. This "other side" is accessible by boat or by walking to Leland's Main Street (M-22), crossing the bridge above the dam, and continuing to the sandy access road behind the south-side buildings. One of these buildings belongs to the Leelanau Historical Society and is used by FPS for equipment storage and net mending. The others are private residences, a vacation rental, and a guest lodge.

Only two of Fishtown's main buildings are used for commercial fishing, down from a high of nineteen commercial fishing shanties, net sheds, and ice houses blanketing the site during the peak fishing period of the 1930s. The fishing fleet has been cut to two tugs, down from its historic high of eight. But working fish tugs still moor alongside the weathered docks.

Fisherman Alan Priest captains the *Janice Sue* and understands just how special Fishtown is. "It's not just a place," he says. "That's my whole world." During the busy summer months, as Priest works on the docks unloading boxes of fish and nets, he takes satisfaction in knowing that thousands of annual visitors also recognize the unique qualities of Fishtown. "I hear people say, 'I've never seen a spot like this, anywhere.'"[4]

FISHTOWN MAPS

Fishtown Site Master Map, 2011 (with identifying numbers for buildings). Historic names are drawn from the original fishermen occupants. With the exception of Louis Steffens, fishermen likely built the shanties themselves. Note: grey-shaded buildings identify those owned by the Fishtown Preservation Society through its 2007 purchase. All other buildings are privately owned.
Map by HopkinsBurns Design Studio, Ann Arbor.

1. Historic Name: **Henry J. Steffens Net Shed**; Common Name: **Tug Stuff** (c. 1926)
2. Historic Name: **Ice House**; Common Name: **Haystacks, Tack and Jibe, Spörck Tileart** (c. 1926)
3. Historic Name: **Porthole Building, North Manitou Ferry Ticket Office**; Common Name: **Manitou Island Transit** (1980)
4. Historic Name: **Porthole Building, North Manitou Ferry Ticket Office**; Common Name: **Manitou Island Transit** (1972)
5. (Two historic shanties now combined into one building) Historic Name: **Claude Kaapke and Eli Firestone Shanty** (c. 1926) and **Peter Nelson and John Maleski Shanty** (1928); Common Name: **Carlson's of Fishtown** or **Carlson's Fishery**
6. Historic Name: **William Harting and Oscar Light Shanty**; Common Name: **The Crib** (c. 1926)
7. Historic Name: **Severt Johnson and Nels Carlson Shanty/Henry J. Steffens Shanty**; Common Name: **Diversions** and **Bead Hut** (1905)
8. Historic Name: **Louis Steffens Shanty**; Common Name: **Village Cheese Shanty** (c. 1959)
9. Historic Name: **Reflections Art Gallery**; Common Name: **Dam Candy Store** (1969)
10. Common Name: **Reflections Art Gallery** and **Leland Beach Company** (1979)
11. Historic Name: **Fisherman's Cove Restaurant**; Common Name: **The Cove** (1967)
12. Common Name: **Falling Waters Lodge** (1966)
13. Historic Name: **North Manitou Ferry and Mail Boat Warehouse**; Common Name: **Nicholas and Susann Lederle House** (1928, additions 1980, 1986, 2012)
14. Historic Name: **William Smith/William Buckler Shanty**; Common Name: **Leelanau Historical Society Shanty** (portions of c. 1908 building moved across river, cut apart and rebuilt with new center section in 1928)
15. Historic Name: **George Cook and Martin Brown Shanty**; Common Name: **The Otherside Vacation Rentals** (1903, remodeled 2000-2001)
16. Historic Name: **Warren Price Shanty**; Common Name: **Hall Shanty** (c. 1918)
17. Common Name: **Storage Shed** (late 1970s)

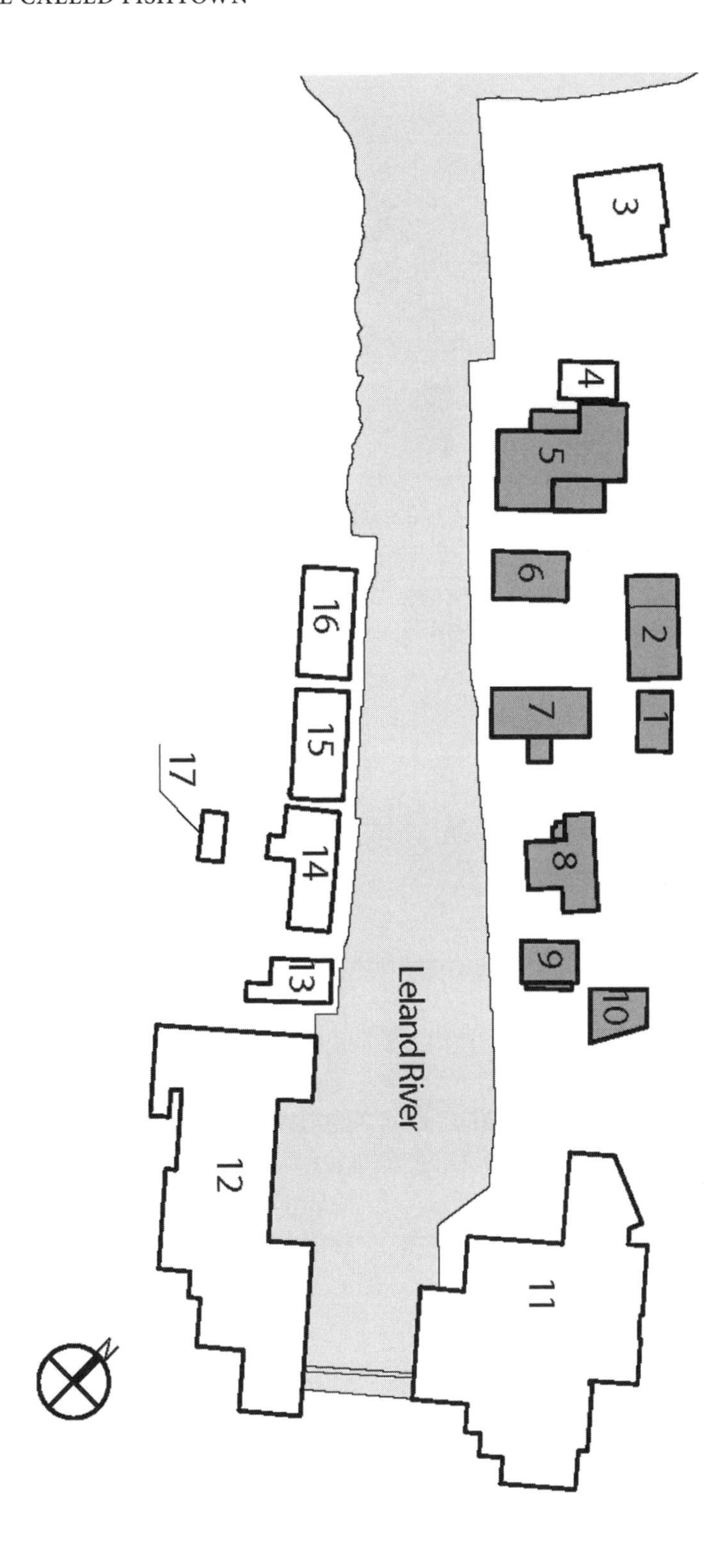

Fishtown Site Master Map, 2011

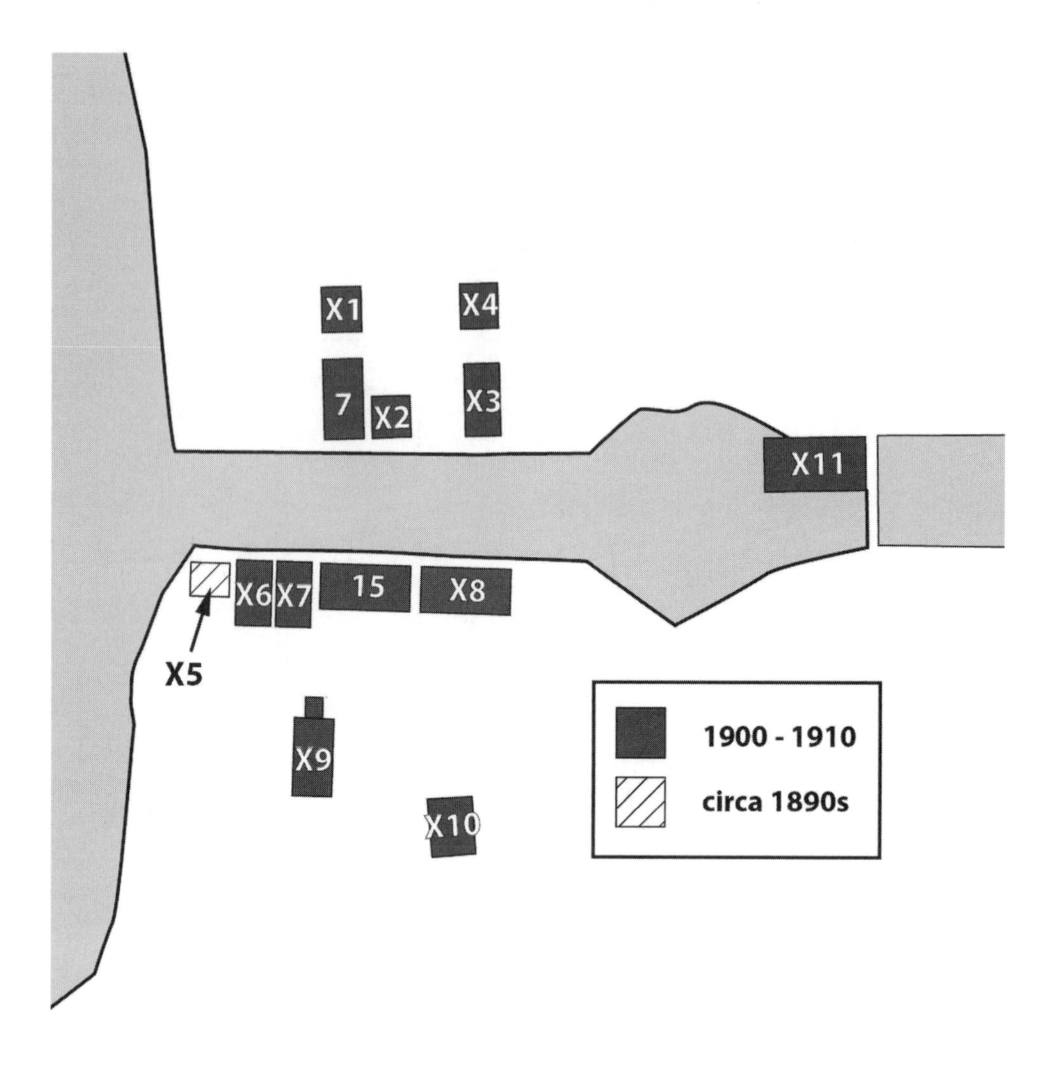

Fishtown's Historic Footprint Timeline Map
1900-1910

(Buildings with an "X" no longer exist). MAP BY HOPKINSBURNS DESIGN STUDIO, ANN ARBOR

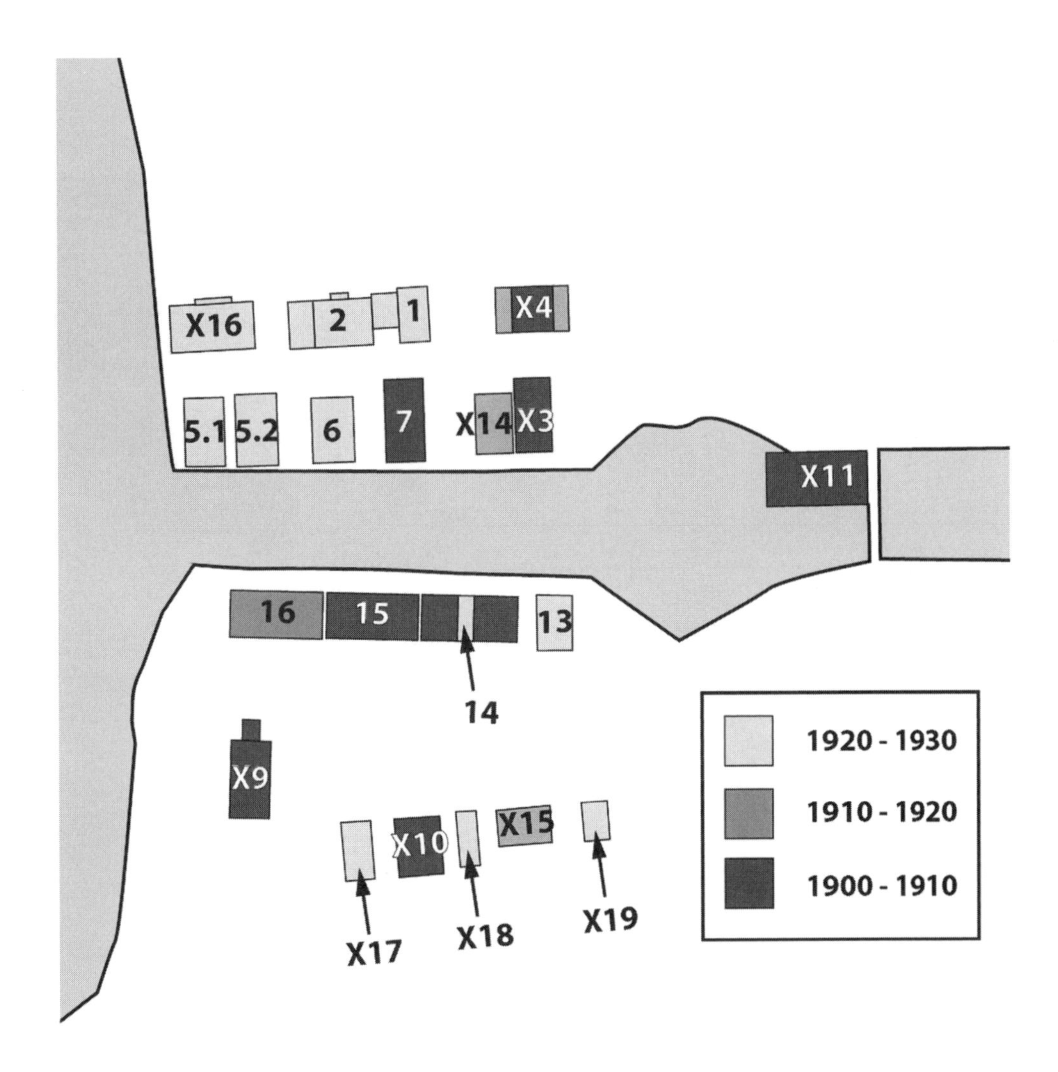

Fishtown's Historic Footprint Timeline Map
1920s-1930s

(Buildings with an "X" no longer exist). MAP BY HOPKINSBURNS DESIGN STUDIO, ANN ARBOR

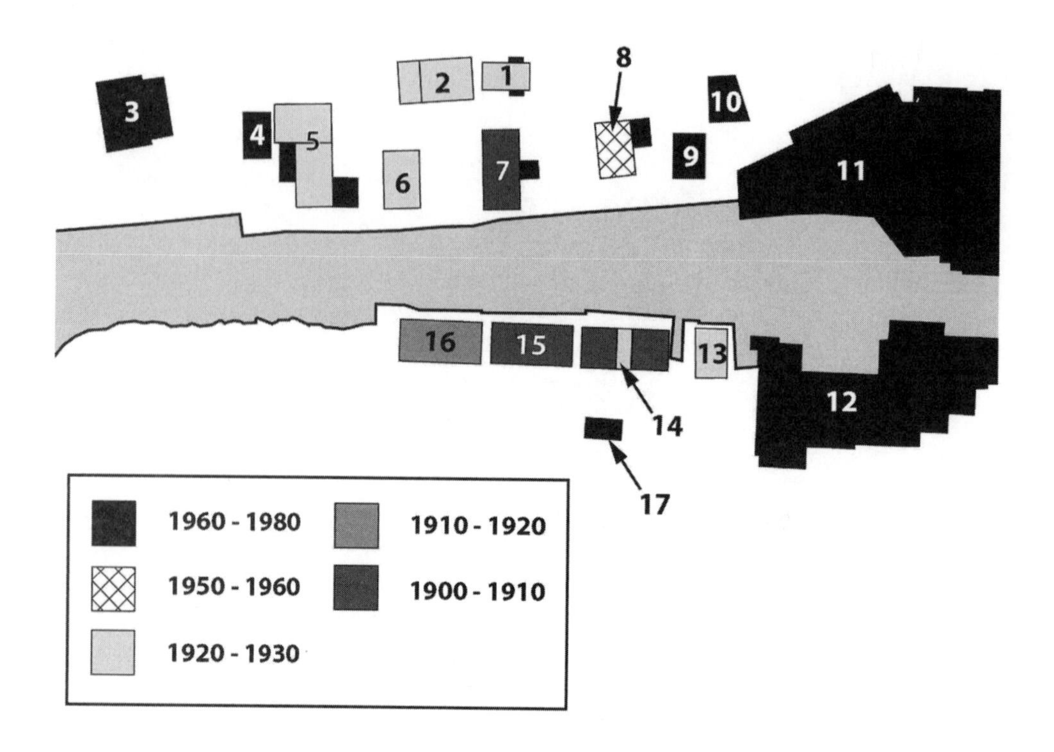

Fishtown's Historic Footprint Timeline Map
1960-1980

MAP BY HOPKINSBURNS DESIGN STUDIO, ANN ARBOR

Chapter 2

Why Fishtown Matters

"Fishtown's unique collection of fishing shanties is unparalleled in the Great Lakes."[1]
–Alan Moore, 1974

THE HEADLINES ARE STARK: "COMMERCIAL FISHING, ONCE A GREAT LAKES way of life, slips away"; "Commercial Fishing Was Once a Major Industry for Some Communities"; "The Shrinking Commercial Fishing Industry." The storyline describes many former Lake Michigan fishing ports. For Joel Petersen, who began fishing out of Leland about three years ago on the *Joy*, the headlines are personal. Petersen, of Muskegon, is a third-generation commercial fisherman. "It's kind of a lost art, because there are so few of us left," he says. "There used to be guys everywhere—Frankfort, Manistee, Ludington, Grand Haven, Whitehall, Pentwater. Every port had their fishermen. Now they're gone, so everybody's forgotten about it. That generation is gone. The younger generation that never experienced it, they have no idea that it even exists."[2]

Bud Stenberg of Pentwater does remember, even though he hasn't fished since the mid-1960s, when changing state regulations forced him to retire. Like Leland, Pentwater is a tourist destination located on the shores of Lake Michigan. But Pentwater has lost its fishtown. Leland, miraculously, still has one. So fishermen like Stenberg visit Fishtown, bringing with them faded photos and memories of the vanished fish shanties in their hometowns. "Commercial fishermen, who say that fishing is in their blood, love Fishtown," observes Amanda Holmes, who

Adam Kilway and Jerry Vanlandschoot work on the deck of the *Joy*, one of two fish boats now owned by Fishtown Preservation as part of its mission to sustain Fishtown's historic working waterfront, 2010. PHOTO BY MEGGEN WATT PETERSEN

serves as the Fishtown Preservation Society's executive director. "To me, that's a measure that we're doing something right."[3]

For Leland's Bill Carlson, and other long-time commercial fishermen, "It isn't a job, it's a way of life." Like most sons of commercial fishermen, Carlson grew up around the water, boats, and fishing. He earned small change when he was just five years old by filling the net-mending needles. By his teens he had done most of the jobs required of a commercial fisherman. Then he attended college and left Fishtown behind. But when his great-uncle Gordon—fishing partner to Bill's dad—became ill, Bill returned home and spent four more decades in commercial fishing. "For a time, this was the last thing in the world I wanted to do," he told a newspaper reporter in 1977. "Now this whole place feels good to me."[4]

The "feel" of Fishtown helps to define the place. The creak of the docks underfoot. The clean, clear smell of fresh water. The flutter of colored flags topping a discarded anchor buoy. The rush of water tumbling over the Fishtown dam. The chug of engines from charter boats, fish tugs, the Manitou Island ferry. The textures of wood: pilings, shingle, shiplap. The sight of working boats, working fishermen, boxes of fresh-caught fish. The tang of burning maple wood as smoke cures the racks of trout, whitefish, and salmon in the double smokehouse next to Carlson's fishery. The spray of water hosing off the fishery floor. The smell of fish. Questions from tourists: *What kind of fish is that? How many fish do you catch a day? How many fish do you clean a day?* "Fishtown is one of the only places," writes Amanda Holmes, "where we can still see and feel a connection to the long tradition of Great Lakes maritime culture."[5]

There were times when it seemed that this connection would be lost and Fishtown's working waterfront, too, would disappear. Twenty years ago the *Christian Science Monitor* described Fishtown as "an ersatz tableau of what Great Lakes fishing was in its heyday. Its dockside bustle enchants tourists but masks how commercial fishing in Michigan is just a shell of what it once was." Yet Fishtown has sustained just enough fishermen to maintain its character as a *real* fishing village.

Amanda Holmes knows firsthand just how special Fishtown is. Since joining Fishtown Preservation, she has traveled the shorelines of the western Great Lakes to seek out remnants of this once thriving industry. Even among active fisheries, Fishtown's collection of original wood fishing buildings sets it apart. Elsewhere, abandoned fishing structures succumbed to wind and weather or were torn down to make room for new waterfront development. In the surviving fisheries, pole barns or concrete block have often replaced the original wood. In Fishtown, generations of fishermen repeatedly patched their small wooden shanties against the ravages of Mother Nature and kept fishing. When Fishtown Preservation acquired its Fishtown properties in 2007, the organization understood that its mission was to protect a survivor. "We want to keep Fishtown alive not only for this one place," Holmes emphasizes, "but as a way of keeping alive a way of life that once existed around the entire Great Lakes."[6]

The health and vitality of Fishtown's way of life depends on the health and vitality of the fish populations. In 1930 Michigan had the most extensive fishery of any Great Lakes state, engaging 2,237 men in the business and producing 2.5 million pounds of fish annually. Leland was a microcosm of the larger picture, a small operation with just 8 fish rigs employing 18 families, or 72 people. The federal government's 1935 harbor report estimated the value of the Fishtown catch in 1933 at $50,000 and observed, "Although the commerce is small in actual tonnage and value, it is of vital importance to the islanders and to the

Fishtown postcard, c. 1920. COURTESY OF LEELANAU HISTORICAL SOCIETY

fishermen at Leland."[7]

By the 1970s the situation had changed dramatically. Michigan Department of Natural Resources policies favored recreational fishing, restricting state commercial licenses to fish species of low value to sports fishermen, and confining commercial fishermen to gear with minimal by-catch, such as trap nets. The landmark Fox Decision (1979) upheld historic American Indian treaty rights in Lake Michigan and forced many non-Native commercial fishers out of business. "I was put out of business eleven times in ten years, for one reason or another," Bill Carlson recalled. "A lot of it had to do with regulations, some with contamination, some with the fish stocks and the exotic species, and the whole environmental change in the lakes."[8] By the mid-1970s, Fishtown housed just three commercial fishing operations.

The remaining fishermen adapted and re-tooled, as they always have, but numbers plummeted. Today Michigan issues only fifty state licenses for commercial fishing. Tribal licenses, issued separately since the treaty rights litigation, totaled just over one hundred in 2011. A few commercial fishermen still make a good living, but Great Lakes ecosystems are in crisis. Biologists point to a

combination of factors affecti
pollution, global warming, and c
The fish are disappearing, and with
Michigan, there are now just seven state-
representing fifteen licenses. The Fishtown P

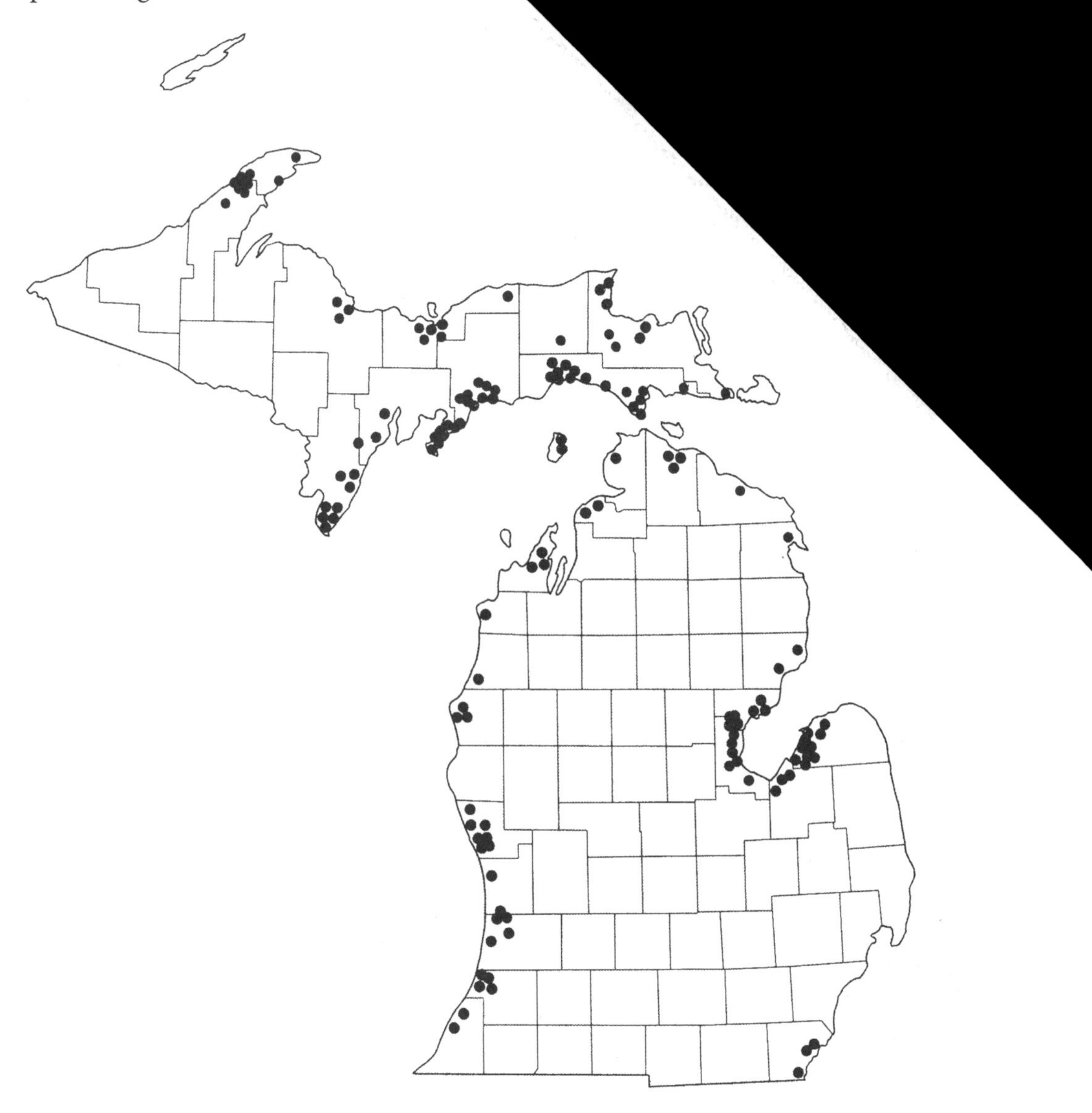

Commercial Fishing Ventures, 1974 data, from the *Atlas of Michigan* (1977), courtesy of Michigan State University Press. Each dot represents the location of a commercial fishing operation. This data includes tribal fishers operating under state license as well as non-Native commercial fishermen.

ystem, but
re organic
ng fishing
ning in the
i's survival.
that would
ommunity
ntown and,
ures, which
mpromise,

ittee listed
. Its report
otographed
ge (despite
came retail
ry modern

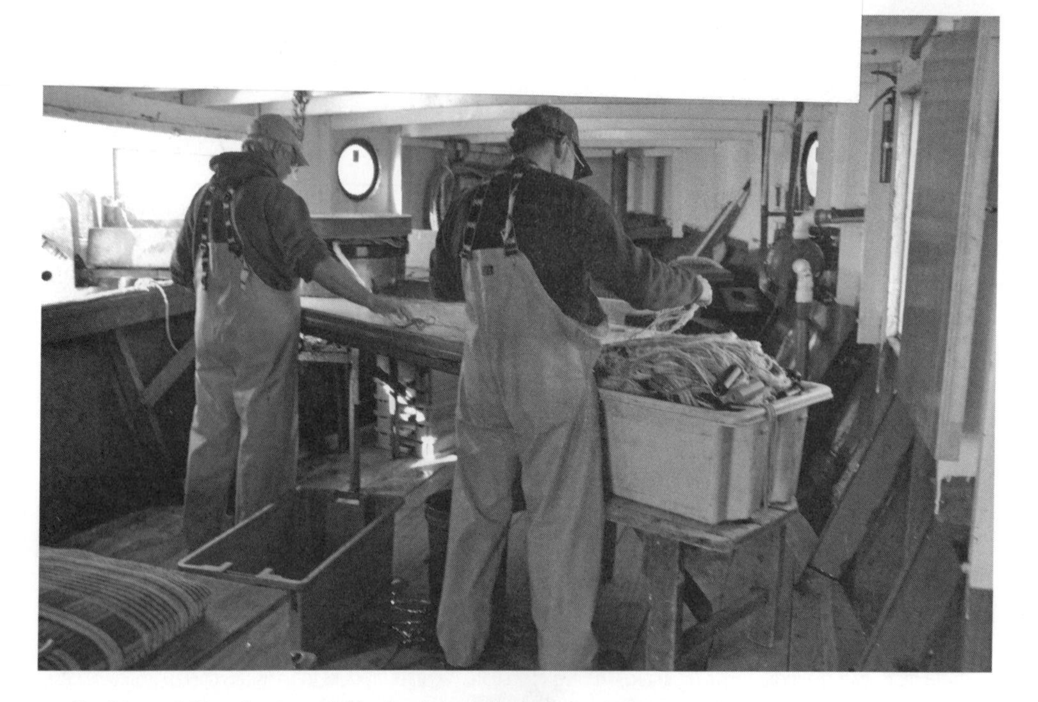

Alan Priest and Albert Gunderson fishing for chubs on the *Janice Sue*, 2010. PHOTO BY MEGGEN WATT PETERSEN

buildings were constructed for the growing tourism market. Completed in the mid-1960s, Falling Waters Lodge and Fisherman's Cove Restaurant anchored Fishtown's eastern boundary. Their massive concrete footprints obliterated some of Fishtown's past. Overgrown slag piles from the nineteenth century iron furnace became building sites. The boat yard—where several generations of fish tugs were built, stored, and repaired—was partially covered by a riverside deck where tourists enjoyed cocktails. And above the site of the former ferry loading dock, guests at Falling Waters fished from their balconies, unaware of the vanished landscape.[11]

Tourism bolstered the local economy, but by the early 1970s some local residents were concerned that the new developments compromised the community's sense of place. The new harbor and marina brought increased pollution and crowds of tourists that changed the character of Fishtown. Leland was beginning to revise its zoning ordinance to more effectively protect its waterfront and small-town atmosphere from unwanted development. Local residents called on Michigan's newly instituted State Historic Preservation Office to move forward with a Leland

This postcard of Fisherman's Cove Restaurant and Falling Waters Lodge, both built in the mid-1960s, promoted a tourist image of Fishtown devoid of shanties and working boats. COURTESY OF LEELANAU HISTORICAL SOCIETY

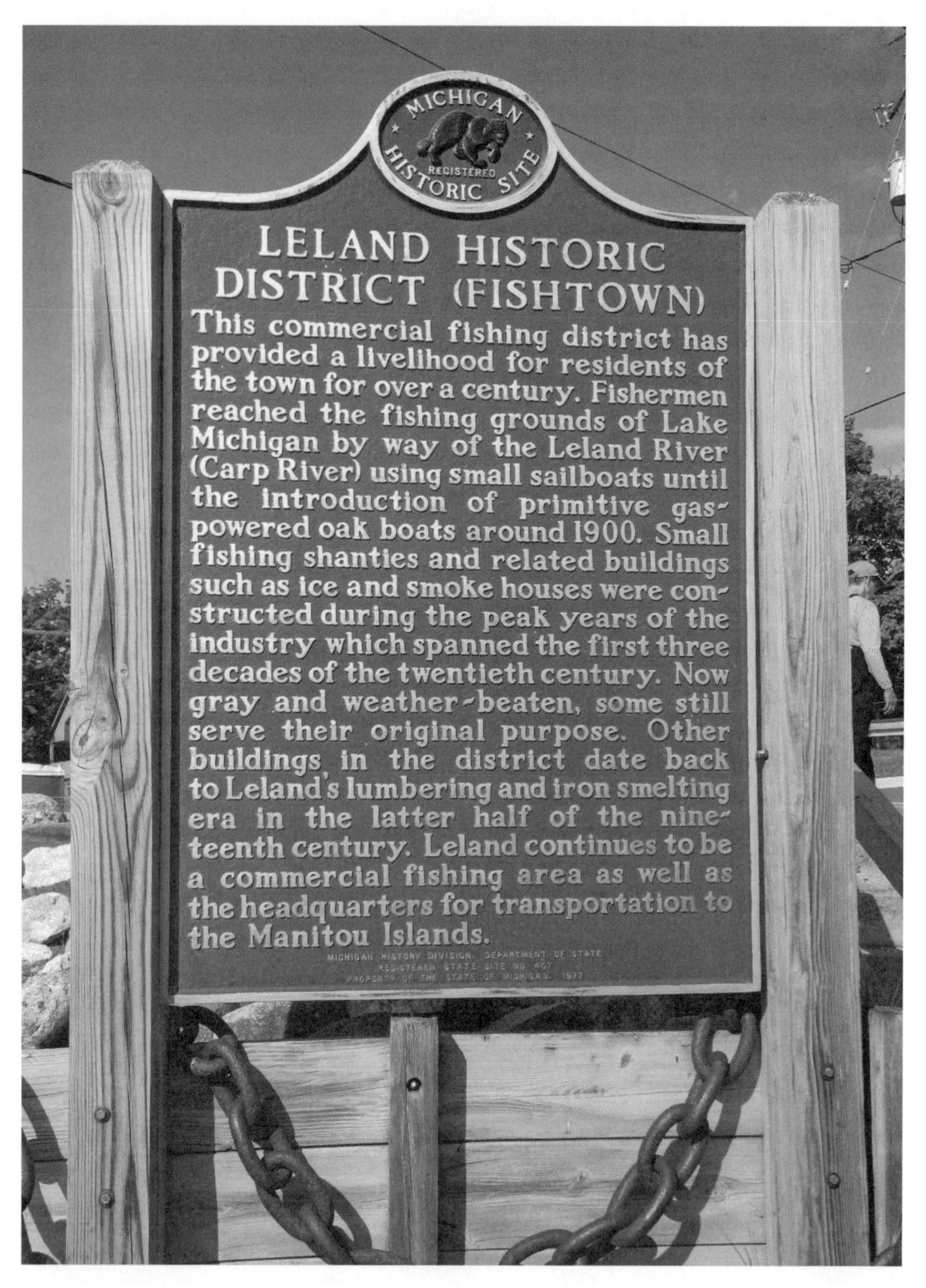

Fishtown received recognition from the Michigan Historical Commission's marker program in 1977. PHOTO BY LAURIE KAY SOMMERS

National Register Historic District nomination, which included Fishtown and Leland's historic downtown. Under provisions of Section 106 of the National Historic Preservation Act (NHPA) of 1966, the historic district designation—even the eligibility for listing—called for comment by the Advisory Council on Historic Preservation on the effect of the federally funded, licensed, or regulated actions at the harbor area. The NHPA created the National Register of Historic Places and the State Historic Preservation Offices as part of a comprehensive plan to preserve the nation's historic and cultural resources. The historic preservation community agreed with local residents that Fishtown was a place that mattered, with listings on the State Register of Historic Sites (1973), the National Register of Historic Places (1975, as part of the Leland Historic District), and the erection of a Michigan Historical Marker (1977). During the same period, University of Michigan master's student Alan Moore completed a study for Michigan Sea Grant on policy and preservation considerations in Michigan's historic fisheries. He concluded that in Leland's unique Fishtown, "a modern commercial fishing operation should be maintained at all costs."[12]

Fishtown is more significant today than it was when first listed on the national register. It is increasingly rare to find working commercial fisheries on the Great Lakes. It is rarer still for a working fishery to be so accessible to the public, operating *in situ*, in historic shanties that have housed generations of commercial fishermen.

A place so vulnerable would not have survived without the local community. These were the people who lived, worked, and played in Fishtown and shared its stories. Place Matters founder Ned Kaufman coined the term "storyscape" to underscore the ways places are used and made meaningful to the community that sustains them. Storyscapes like Fishtown matter not just for their architecture but also for their ability to "convey history, support community memory, and nurture people's attachment to place."[13]

Fishtown Preservation understands the importance of story to its mission. Following the precedent set by the Leelanau Historical Society, FPS has conducted interviews and scanned photos and old scrapbooks not only for facts and images, but also for the stories they contain: of unique personalities, storms and ice on the Big Lake, accidents and close calls, the uncanny skill of old-timers, and enduring attachments to Fishtown.

A recent Fishtown Preservation newsletter highlighted one particular story as a metaphor for why Fishtown matters. The story involves the tug *Good Will*, the last wooden fish boat built for Leland's fishery. The boat owed its existence to the good will of the townspeople, who helped pay for its construction after a terrible accident destroyed its predecessor (see the Fishing Generations chapter for more about this story). This same "good will" allowed the boat's owner, Leland fisherman Pete Carlson, to continue commercial fishing.

One stormy night in December 1963, Pete was motoring across the Manitou Passage in a blinding snowstorm. On board was his uncle and fishing partner, Gordon Carlson, a mother lode of fish, and sodden gill nets from their recent lift. Fish tugs had long shared the historic shipping channel with massive freighters that plied the Great Lakes. Even in the age before radar, collisions were rare. But on this day, the big freighter churning through the stormy seas never saw the smaller gill-netter. A big wash from the freighter's bow buffeted the *Good Will* just before she was hit on the stern and spun completely around, knocking the two fishermen to the floor. The force of the blow cracked the boat's ribs and

A small boy, Allen Northcutt, fishes from a tug as the *Good Will* motors into Fishtown, c. 1950.
COURTESY OF BILL CARLSON

Carl Jonathan Hanpeter shows that "This Place Matters," a special designation from the Place Matters campaign of the National Trust for Historic Preservation, 2010. PHOTO BY AMANDA HOLMES

punched a hole in her hull. What happened next might be called "the miracle of the fish," turning potential tragedy into good news. As fisherman Jim VerSnyder later heard from Pete, "If they hadn't had a thousand pounds of fish in the boat at that time, they might have sunk. Because when the freighter hit them on the stern, the fish shifted to the other side; it tipped the boat on her side and pulled the hole out of the water. So they came in kind of running on their side. So it was lucky to have that many fish!" The Fishtown Preservation Society, now writing its own chapter in Fishtown's story, put it this way in a recent newsletter: "Pete and Gordon nursed their listing tug back to Leland, saving the *Good Will*, their livelihood—and possibly their lives—through the literal weight of that day's good news. They not only made it home in the *Good Will*, they also patched her up and kept fishing. That's what we do in Fishtown—we patch and keep fishing."[14]

For Fishtown Preservation, the good fortune of the *Good Will* encapsulates the Fishtown story. Against the odds, Fishtown is still standing. Against the odds, Fishtown continues its proud commercial fishing legacy in the newly restored tugs *Janice Sue* and *Joy*. Fishtown and its stories matter.[15]

Bela Hubbard's sketch of the Ottawa town located north of the mouth of the Carp River, 1838, shows two typical cabins, a vegetable drying rack, and a Manitou pole integral to Ottawa religious beliefs. IMAGE COURTESY OF BELA HUBBARD PAPERS, BENTLEY HISTORICAL LIBRARY, UNIVERSITY OF MICHIGAN.[3]

Chapter 3

Before Fishtown

> "Leland of 1864 [was] a village of shacks clustered near the river and Lake Michigan, with a population of about 150, chiefly lumbermen and woodcutters. Dances were held frequently, and fights occurred at every dance."[1] –Henry J. Barnard, 1927

When University of Michigan professor Alexander Winchell surveyed the Grand Traverse region's geological and industrial resources in 1866, he found vistas of breathtaking beauty: "Carp lake is a beautiful sheet of pure water, resting in the bosom of the hills, which, with their rounded forest-covered forms, furnish it a setting of surpassing loveliness. Carp river, the outlet of Carp lake, discharges a body of water nearly as large, and having a fall of 5 or 6 feet [at the dam] affords another admirable water power. This river is not over a half mile in length."[2]

The Carp (now Leland) River shaped the evolution of human activity along its banks. Estuaries and lakeshore locations of the Grand Traverse region, including the Leelanau Peninsula, were used by the Grand Traverse Ottawa and Chippewa to gain access to the rich fishing grounds of Lake Michigan and Grand Traverse Bay. The area north of the Carp River mouth was the site of an Ottawa village called (with various phonetic spellings) "Shamagobeg." Anthropologist James McClurken suggests that this was the oldest and largest permanent village of the Grand Traverse bands of Ottawa. The 1839 census lists 239 people, "nearly as many as the second and third largest Grand Traverse villages combined." Just

Leland dam, c. 1900, presumably on or near the site of Manseau's original dam, with the abandoned Leland Lumber Company sawmill to the left. COURTESY OF LEELANAU HISTORICAL SOCIETY

a year before this census, young Bela Hubbard and state geologist Douglass Houghton had arrived by boat at Carp River as part of the Michigan Geological Survey. Hubbard's extraordinary drawing of Shamagobeg is the earliest visual documentation of buildings and structures near the river's mouth.[4]

The French Canadian millwright Antoine Manseau would have seen the town of Shamagobeg while scouting the mouth of the Carp River in the late 1840s for a possible sawmill site. The river ran swift and free, falling 15 feet from its source in Carp Lake (present-day Lake Leelanau) to its outlet in the Big Lake. Undeterred by the presence of Ottawa and Ojibwa families, Manseau and his son, Antoine Jr., returned in the summer of 1853 to build the first dam and sawmill on the Carp River. Most early histories state that the Indians of Shamagobeg left soon after the arrival of the early white settlers, believing all the land had been sold to whites.[5]

The Indians may have abandoned their town site, but they relocated to other areas of Leelanau County and still frequented the new village of Leland. New York-born Henry J. Barnard arrived in Leland in June 1864, and recalled that "Indians were numerous." Barnard and his young family had come to join his father-in-law, R. A. Stanbrough, who had a shingle mill on the south bank of the river in what is now Fishtown. "A man named Thies had a small lumber and grist mill on the north bank," he remembered, "and a fisherman named Mike Daly plied his trade on Lake Michigan with a sailboat. The first bridge had a gap in the middle to allow boats to pass through, and those desiring to cross had to lay a plank over the gap."[6]

Leland's pioneering white settlers were primarily lumbermen and woodcutters. Wooding, or providing cordwood fuel for the steamships traversing the Manitou Passage, was the engine of commerce throughout the Grand Traverse region. Wooding stations on the nearby Manitou Islands developed as early as the 1840s, while mainland communities such as Leland followed suit in the 1850s and 1860s. By 1867 Leland's population had grown to two hundred. In addition to three

Tugs and scows hauling cordwood down the upper Carp River, 1863. COURTESY OF LEELANAU HISTORICAL SOCIETY

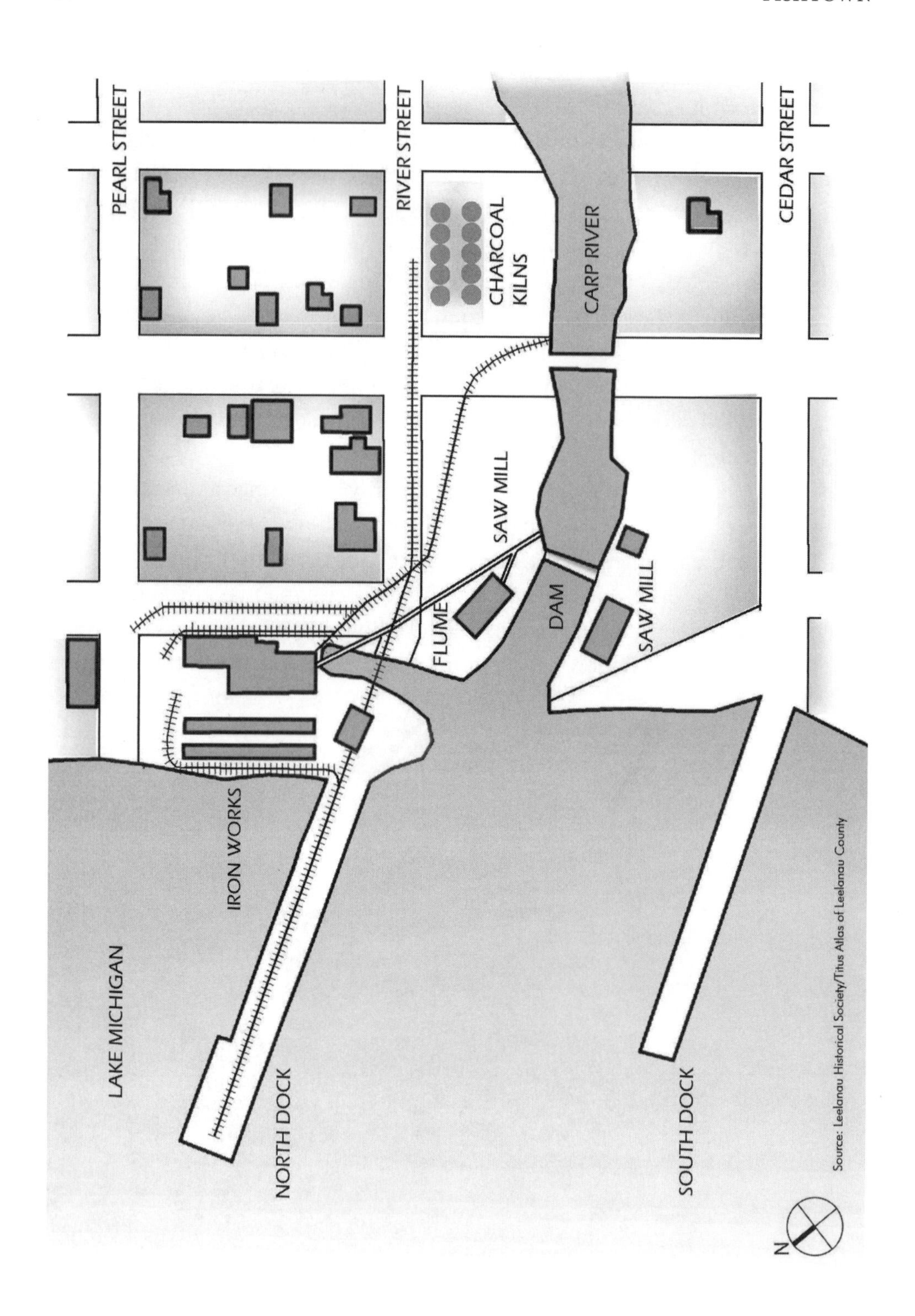
PEARL STREET
RIVER STREET
CEDAR STREET
CHARCOAL KILNS
CARP RIVER
SAW MILL
FLUME
DAM
SAW MILL
IRON WORKS
LAKE MICHIGAN
NORTH DOCK
SOUTH DOCK
N
Source: Leelanau Historical Society/Titus Atlas of Leelanau County

shipping docks, the *Michigan State Gazetteer* reported the nascent village had "three stores, a saw and grist mill, a stave and heading manufactory, two hotels or boarding houses, shoe and blacksmith shops, a physician and grocer." Leland became the principal business point for Centerville Township. When, in March, 1867, Manseau's dam was swept away, commerce was devastated. The new, higher dam raised the water level 12 feet. What had been three connected lakes upstream became one large lake, Carp Lake, now navigable to Cedar and providing access to vast interior tracks of hardwood. A report of the time extolled the virtues of the little town: "Located, as it is, between Carp Lake and Lake Michigan, with Carp River connecting, fine facilities are offered and improved, for converting the stately forests surrounding Carp Lake, into wood, staves, etc., by means of tugs and scows." Sawmills flanked the river, their saws buzzing through the massive timbers that floated downstream to Lake Michigan. Lumbering dominated the economy until 1870, when the Leland Iron Works fired up its blast furnace and changed the face of Leland.[7]

The river's clear shallow waters today belie its past as a conduit for industry. "The charming little village in northwest Michigan" now touted on Leland's official website was, in the late nineteenth century, a grimy company town. Its favorable location had caught the eye of a group of Detroit entrepreneurs who founded the Leland Lake Superior Iron Company, an enterprise that existed under various names and ownership from 1870 to 1884. Leland seemed ideal for the new venture due to waterpower provided by the river and dam, a vast supply of timber, and proximity to Lake Michigan where ships transported raw iron ore from the Upper Peninsula and then carried ingots of pig iron to steel mills on the Great Lakes. The iron company dominated the landscape and owned large portions of Leland, including what is now the north side of Fishtown.[8]

For the few fishermen then working out of the river, it must have been a relief to escape to the fresh, windblown expanse of Lake Michigan, away from the grit and grinding noise of the smelter and charcoal ovens. A poem published in the *Leelanau Enterprise* captures the sudden feeling of quiet relief of "Leland in November" after the shipping lanes had closed for the season:

(OPPOSITE PAGE) Map of Leland, 1870s, showing industrial use of the land adjacent to the river. The south-side sawmill locations are inferred from Henry J. Barnard's recollections of Leland in 1864. MAP BY JOHNSON HILL LAND ETHICS STUDIO, ANN ARBOR

No boats, no smoking kiln,
No whirling, buzzing mill,
No fire from out the stack,
No charcoal, dust nor slack,
No crushing of more ore,
No business any more,
No smelt, no blast
All's quiet at last—
Leland in November.[9]

The iron company's boom proved short-lived. When, in 1884, the last in a succession of owners declared bankruptcy, Leland's economy was devastated. As the 1887 federal harbor report observed, "At present its commerce has dwindled to almost nothing, its present appearance is that of a town that has been going backwards for years." The newly formed Leland Lumber Company bought the property for its sawmill, but as loggers clear-cut inland forests, the lumber company also went into receivership. By 1900 the company's large pier had been swept away in a storm and the blast furnace dismantled. Only the abandoned sawmill still stood at river's edge, and then that, too, was torn down. When legal issues surrounding the property finally were resolved, a large section of Leland,

View of Leland, showing the blast furnace, charcoal kilns, and stacks of cordwood along the upper river above Fishtown, 1870s. COURTESY OF LEELANAU HISTORICAL SOCIETY

This image from the late 1800s shows the Leland Lumber Company sawmill to the center left. A fish shanty is situated to the far right, with at least five Mackinaw boats moored in the river. The top of a pile driver (used to set and pull stakes in a pound net) appears to the right of the sawmill. The building to the left of the sawmill also may be a fish shanty. COURTESY OF LEELANAU HISTORICAL SOCIETY

including what is now Fishtown, became available for other uses. In March 1900, the *Leelanau Enterprise* enthusiastically proclaimed, "Alive again, Leland promises to be one of the leading towns of the County."[10]

Leland in 1900, however, was not yet an inviting place. Although its legacy as resorter paradise, artist colony, and fishing village was about to begin in earnest, the landscape had not yet recovered from three decades of industrial abuse. The writer Karl Detzer recalled that the river above the dam "had a channel that did not look more than 10 feet wide, with flats on both sides where thousands of pine and hemlock stumps stuck out of the water. It was not a pretty stream. Nor was Leland a pretty town. The sidewalks were wooden, of course, with many planks missing. The town had two saloons and how the customers got home over those sidewalks always was a mystery. They didn't dare walk in the grassy street because that was where the cows walked, and cows were very untidy."[11]

Fishtown at the turn of the twentieth century was not a pretty place either. The site was cloaked in sawdust and iron slag (today known popularly as Leland bluestone). Lake Michigan's fishing banks teemed with whitefish, herring, sturgeon, and lake trout, but commercial fishing was still a modest enterprise, reflecting its recent past as a stepchild to sawmills and the iron works. Early resorter Joseph Littell recalled, "Remnants of a few mills lay idle and abandoned beside their empty lumber yards, and the silence was broken only by the whistle of the incoming [steamer] *Tiger* from Fouch, and the shrill whistle of John Peter's portable sawmill which lingered along the river to saw the last log at Leland. The old black piles still stood above the lashing waves of Lake Michigan where once the docks had been, each well worn stump providing a perch for a sea gull."[12]

Chapter 4

Seasons of a Fishery

"Eight fish boats hailing from Leland will ply the waters of Lake Michigan when the trout season re-opens early in November, ready to harvest their share of the finny crop."[1]

– *Leelanau Enterprise*, 1938

FISHTOWN AS WE KNOW IT TODAY BEGAN TO TAKE SHAPE AFTER THE TURN of the twentieth century. Early resorter Joseph Littell remembered those days: "A few fishermen, the Cooks, the Prices, and the Clausses, put out from the mouth of the river below the dam and its abandoned sawmill to their fishing ground in Lake Michigan in sailboats. There was not a power boat among the fishing fleet. And when the nets had yielded their haul and the boats had returned to the fish houses in the river, the campers [pioneering summer resorters] could and did obtain a supply of fresh white fish and lake trout—at five cents a pound."[2]

One of the earliest photos of Fishtown shows three shanties clinging to the sandy bank near the south side of the river's mouth, with Mackinaw boats moored dockside. The blackened piles of the ruined Leland Lumber Company dock appear to the far right, remnants of a once mighty pier that washed away with shoving ice and fierce northerly gales. The buildings to the rear are likely an ice house and net shed, with net drying reels located to the east, shielded by the buildings from Lake Michigan storms. These three zones of activity—river with boats and docks, shanties, and ice houses and net sheds to the rear—came to typify Fishtown's topography.

Unloading the day's catch at Fishtown, early 1900s. COURTESY OF LEELANAU HISTORICAL SOCIETY

Similar fishing villages developed throughout the Great Lakes, each with a common need for architecture and landscape that supported work on lake and shore. Each also had unique characteristics. For Fishtown, the defining geographic feature was the river, a narrow ribbon of water that fostered an intimate clustered village of shanties, ice houses, and net sheds. Fishtown's activity also extended inland and out into Lake Michigan. Local farms provided the space needed for drying and repairing the large nets used in the pound net fishery. Nearby Lake Leelanau was the site of the annual winter ice harvest, when men from the community cut blocks of ice to fill the ice houses for the coming season. The landscape also connected with the waterscape. The river provided the essential conduit to the all-important fishing grounds.[3]

Fishtown continued to develop along "the fish creek," as the river was termed by local residents. The majority of shanties, sheds, and ice houses were constructed between 1903 and 1928. The peak fishing period occurred from the 1920s to the early 1940s. Third-generation fisherman Lester "Pete" Carlson, who began fishing in 1926, recalled, "It was quite an industry back in those days. Directly

This view of Fishtown's south side, c. 1900, shows the emerging template of boats, buildings, and structures on the landscape. None of these buildings exists today.
COURTESY OF LEELANAU HISTORICAL SOCIETY

there were twenty-six families depending on commercial fishermen. That was directly. Indirectly, there was quite a few more, such as those that took the fish to the train, the ice harvest crew, and different families like that."[4]

Traveling salesmen from Chicago, Milwaukee, and Two Rivers, Wisconsin provided most of the equipment used by the fishery. The fishermen carved buoys out of cedar posts, cast their own leads, and built their own nets. Fishtown's commercial fishermen used three techniques common to the Great Lakes fishery: hook and line (a deepwater technique for trout adapted to fresh water by Scandinavian immigrants and also used by American Indians); the shallower water impoundments called pound nets (a Scots' invention introduced into the Great Lakes in the 1830s); and gill nets, a deepwater innovation of Indian fishermen. Trap nets, another impoundment technique, did not enter the Leland fishery until the 1980s, although they were used elsewhere in the Great Lakes as early as the 1850s.[5]

Initially whitefish was the most prized catch, but by the peak fishing period trout was the "money fish." As Roy Buckler explained, "Chubs weren't worth anything back then, you got a dime a pound. I used to fish whitefish in the summertime with pound nets when you couldn't catch trout." Trout fetched up to seven cents

Fishtown, early 1930s, during its peak, with eight gas-powered fish tugs lining the river. PHOTO BY ERHARDT PETERS, COURTESY OF LEELANAU HISTORICAL SOCIETY

Pound nets are set in shallow waters closer to shore. On the right is the leader, a long straight net that leads the fish through the central "heart," which in turn funnels the fish into the "crib" or "pound," where they are trapped.
DRAWING BY L. KUMLIEN OF A POUND NET OFF INGERSOLL'S FISHERY, GREEN BAY, WISCONSIN. COURTESY OF NATIONAL OCEANIC AND ATMOSPHERIC AGENCY'S HISTORIC FISHERIES COLLECTION (NOAA), NOAA NATIONAL MARINE FISHERIES SERVICE

per pound in 1908, while Buckler recalled that during the peak fishing period, "fifteen to eighteen cents was the going price for trout and whitefish."[6]

The life of commercial fishermen cycled with the seasonal round of activities, shaped by weather, increasing government regulation, and the fluctuating fish populations. Pete Carlson and Roy Buckler remembered, "We didn't fish too much after the first of January; we were just about all done then. The boats were all pulled up on the beach." When all the boats were in, the fishermen celebrated by taking a bucket to the nearby Mercantile, filling it with drink, and passing the cup.[7]

In January and February, Buckler and Carlson recalled how "we used to cut our ice off of Lake Leelanau every winter, store it in an ice house, cover it with sawdust, and hope it would last through the summer. Which it always did." Winter was also the time to hunker down next to the pot-bellied stove in the shanty to repair and ready equipment for the coming year. As spring approached and Lake Michigan ice began to thaw, men prepared the boats to set the first gill nets of the season. The first set typically occurred in March or April. "One year," recalled Roy Buckler, "I guess it was the fifth of May before the ice left so we could get out and set nets. It was only an eighth of a mile of ice out here. We

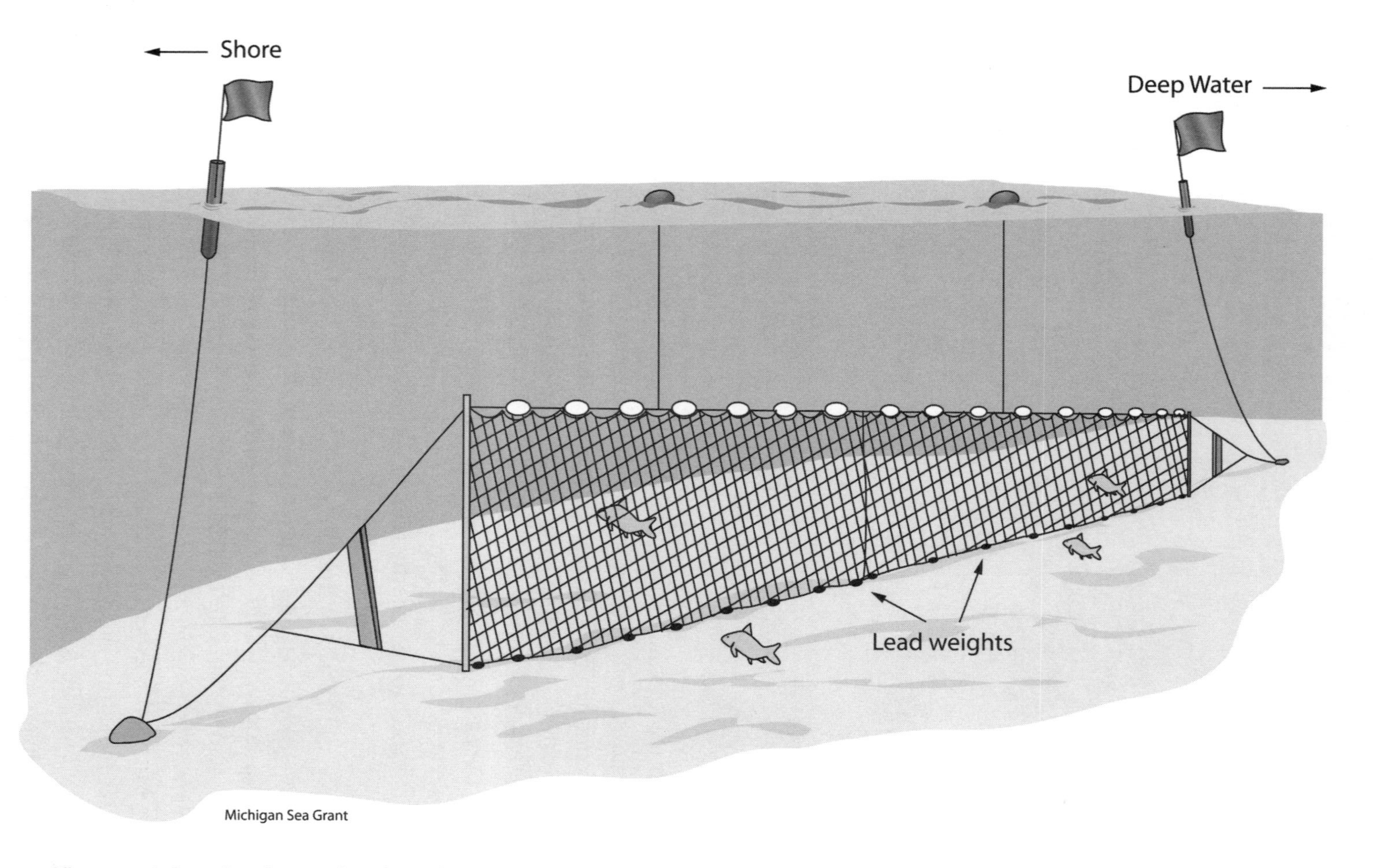

Gill nets contain floats along the top and weights on the bottom that secure the net in the shape of a fence. Fish too large to swim through the mesh are caught by the gills once in the net. Gill nets range from 100 to 400 feet in length and typically are strung together in "gangs." IMAGE AND TEXT COURTESY OF MICHIGAN SEA GRANT'S WEBSITE, KNOW YOUR NETS

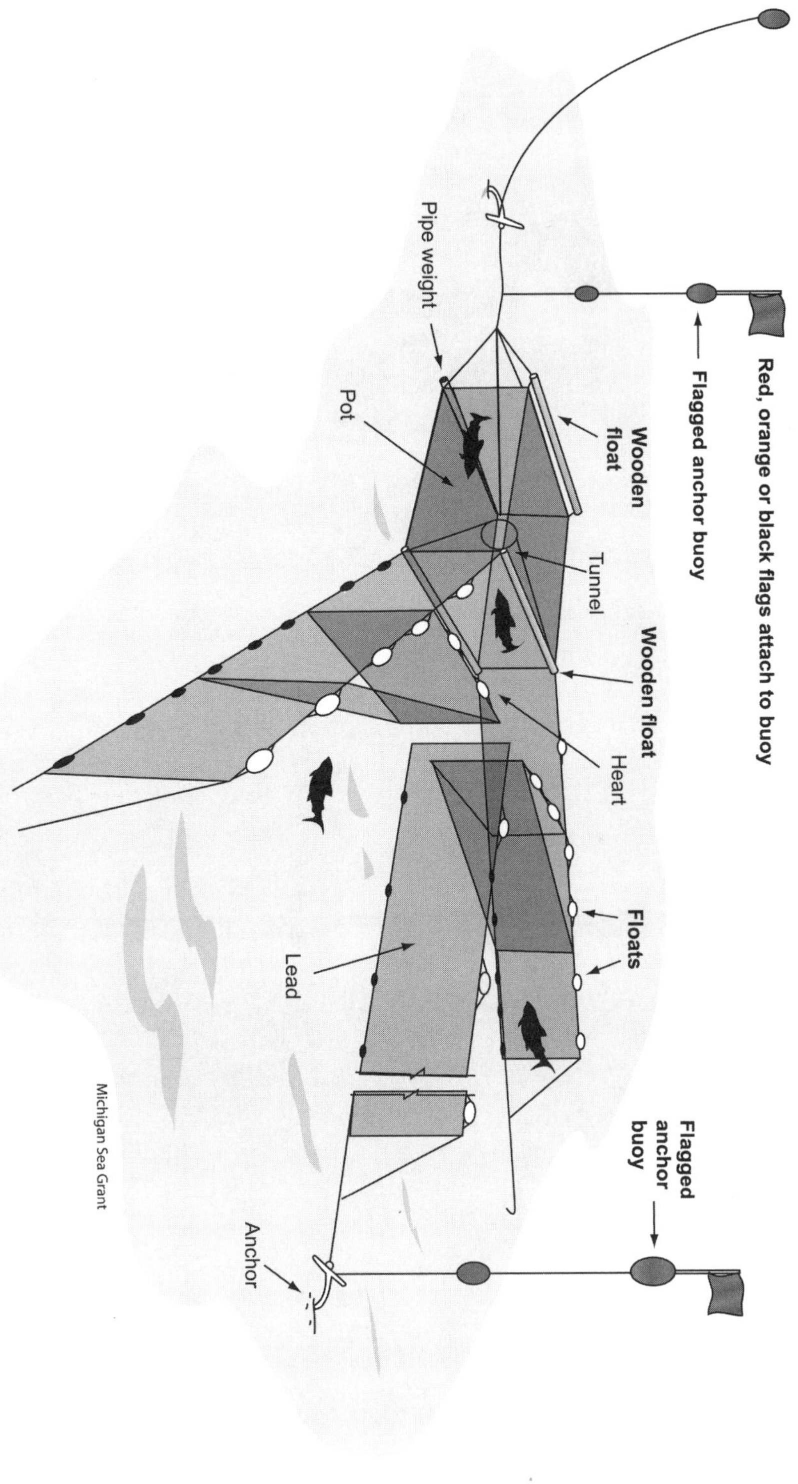

Trap nets are long lead nets that divert fish into an enclosure (heart) and through a tunnel into a pot, where they are trapped. The lead, made of stretched mesh, extends just over 1,000 feet in length. IMAGE AND TEXT COURTESY OF MICHIGAN SEA GRANT'S WEBSITE, KNOW YOUR NETS

Boats dry docked for winter in the Fishtown boat yard, c. 1930. PHOTO BY ERHARDT PETERS, COURTESY OF LEELANAU HISTORICAL SOCIETY

Fishermen lifting their summertime catch from a pound net., c. 1930. PHOTO BY ERHARDT PETERS, COURTESY OF LEELANAU HISTORICAL SOCIETY

dynamited, did everything, but it was right on the bottom of the lake. You'd sit there and look at that strip of ice, and you couldn't do anything about it."[8]

During the summer, in addition to the regular round of setting and lifting gill nets, the fishermen also set pound nets for a lucrative haul of whitefish. "June and July," Buckler remembered, "that's when the fish were there, and we just went down and lifted it every day. Whitefish moved up on the shore to feed in shallow water. Used to have four and five pound nets around the beach and down at the bay. When it came August, that was about the end of it."[9]

In the fall, fishermen collected spawn under special Department of Conservation permits. Beginning in the 1920s, the state closed certain areas of Lake Michigan during fall spawning periods to protect the lake trout fishery. The fishermen carefully removed the eggs from the female trout, and the collected spawn was shipped to state fish hatcheries. At the end of the closed season, reef fishing began. Bruce Price remembered stories of reef fishing from his grandmother, Emma Price: "Most of the trout would be gone, but there'd be a lot of late spawners there. And they were big trout. That's what they always called the reef fishing. You could get big trout, tons and tons of fish. It would be a big thing for all the women to go down. Everybody'd see the catch, see how big they were."[10]

Agents from Chicago typically contracted for fish in the fall, the period of the "big lift," and then negotiated with fishermen throughout the rest of the season according to market prices. The market later expanded to New York, Detroit, and Grand Haven. At one time Warren Price and George Cook, both Leland fishermen, worked as fish buyers. Bruce Price remembers that during his boyhood in the early 1940s, "the big one was George Cook. I think he just represented a company. At Christmas they would come around with pickled herring, sometimes salted herring. As your present at Christmastime from the fish buyers."[11]

During the fishing season, daily activities encompassed both lake work and shore work. Every morning until ice made the lake impassible, fishermen and their wives would rise before dawn. As remembered by Sally Middleton Steffens (daughter of early Northport fisherman Edwin Middleton), "[The women] would prepare a hearty breakfast while the fishermen stoked the boiler on the tug to get up steam [during the early era of steam-powered engines]."[12]

After breakfast the fishermen gathered in Fishtown to decide, as Roy Buckler put it, "whether or not it's fit to go on the water and lift nets." Fishermen like Buckler were adept at reading weather signs. "You go down, and you look at it. You try to get a weather report, which you don't believe, and talk it over with the crew. Of course, in later years, we all had radios and got the weather report from the Coast Guard. You believed all you wanted to of it. If you thought you could still go out and lift, well, you went anyway. Of course, sometimes the weather report was right, and you didn't get your work done. You had to go back and finish the next day."[13]

A fishermen's day was unpredictable. "We would many times leave the harbor at four-thirty, five o'clock in the morning," Pete Carlson recalled. "Sometimes we'd get back before dark, and sometimes it was after dark. Sometimes we wouldn't

Oscar Price always laid claim to the largest trout ever brought into Leland's port. COURTESY OF LEELANAU HISTORICAL SOCIETY

get through work until eleven, twelve o
day. But the difference is—those days and t
faster boats, of course. They do things so much
lift just about as many nets as we used to." The me
nets and cleaned them on the boat. Once back at the sh
tagged their catch before shipment to the railhead. Except
there was little fresh market retail. Instead, as Roy Buckler rec
fish "were hauled out and shipped on the railroad six days a week
have refrigeration, automatic ice makers, and all fish that are shipped
hauled with trucks." After the fish were packed, "a portion of the nets had
dried every day when you come in." Today's nets are built of nylon or monofil
ment, but throughout Buckler's youth everything was fashioned from cotton twine requiring diligent maintenance. "So you spent half your day on the water," Buckler explained, "and half a day reeling and fixing nets."[14]

The work was hard and often dangerous, but fishermen answered to no one but themselves. They made enough money to feed their families and pay the bills, even during the Depression years. Henry Steffens recalled the best lifts of his long career were during the early 1940s. "In two successive days," he recounted

Oscar Price, right, and unknown fisherman spreading nets on the net reel, c. 1930. PHOTO BY ERHARDT PETERS, COURTESY OF LEELANAU HISTORICAL SOCIETY

rout each day, all of
ame disaster: the sea
e trout and whitefish,
s Michigan and Huron
ere over. The *Leelanau*
ng Very Low at Leland"
d lull in fishing success.
reels on the south side
nets, painting boats, and
e 1940s when the fishing
n whitefish and lake trout,"
the fish, which means the
trout were the bread and

mendous lifts of trout were
a di... adily for the past ten years,"
the *Leelanau Enterprise* ... menominees [also known as
pilot fish], perch and a few others now provide ... arly all the tonnage, and the tonnage most of the time is pretty light." By 1958 the fishing fleet was cut in half, with just four boats moored in the river. Two of these were new steel-hulled fish tugs that were christened with great hopes for the future. But the coming decade brought daunting setbacks from which commercial fishermen in Lake Michigan would never truly recover: a botulism scare based in Grand Haven that devastated the fresh market, contaminants like PCBs, a plummeting chub population coinciding with the explosion of alewives, and a series of restrictive Department of Natural Resources (DNR) regulations that stacked the deck against commercial fishermen. "We continually adjusted our fishing to what was available," remembered Bill Carlson, "until the DNR limited us to what we could fish—limited the number of fishermen who could participate, limited the areas and the type of gear we could fish."[16]

When the DNR began managing the lakes for the sports fishery, commercial fishermen had increasingly limited options to remain in business. By 1975 only three commercial fishing operations remained in Fishtown. The *Grand Rapids Press* ran the headline, "Requiem in Leland," reporting fishermen's fears that it was an "unfair deal" and that they were being regulated out of business. Perhaps the most damaging DNR policy was implemented in April 1975. Concerned

Fishtown, early 1950s, as overfishing and the lamprey devastation take their toll. PHOTO BY MIKE BROWN

about overfishing of the chub fishery, the DNR banned chub fishing to all but a limited number of research licenses distributed by lottery. By sheer luck, Bill Carlson won the lottery for the fishing zones that included Fishtown. At that time Leland's remaining commercial fishermen were all part of multi-generational family operations: fourth-generation Bill Carlson, third-generation Terry Buckler, and third-generation Ross Lang, who, with his father Fred, had arrived in Leland in 1968 via the Upper Peninsula and Alpena. To keep the fishing brotherhood going, Carlson vowed, "We'll have to all work together, pooling our nets and boats and using the one permit to bring in income." The three collaborated to form the Leland Fish Company and launched a bold experiment to preserve their livelihoods and way of life.[17]

With federal funds from the Michigan Sea Grant program, they implemented a pioneering application of the purse seine, a large round net, cinched together at the top with purse lines, never before used on the Great Lakes. The Leland fishermen thought this technology would work with whitefish in the northern waters of Lake Michigan. "We built a boat," the *Argo*, remembered Bill Carlson, "and the purse seine worked very, very well. It's no longer used anymore, but that's because the waters that we fished it in are Indian waters, and they have exclusive rights to those areas." With the end of the *Argo*, Terry Buckler eventually quit fishing. Ross Lang switched to trap nets until his tragic drowning in 1998. Bill fished for chub from his boat, the *Janice Sue*, and devised a new vision for the future. The Carlsons shifted their business toward retail and processing, expanding their product line and taking advantage of the booming tourism economy.[18]

Fishtown has become more a tourist attraction and historic site than the fishery of old. Marilyn Steffens Stallman, daughter and wife to Leland fishermen, witnessed this transformation. The Fishtown of her youth was "very rough looking. It was pretty, but in the winter months it was very bare and cold, the wind blowing off the lake. Big icebergs down there on the lake, which we don't get any more. But Fishtown was a wonderful place to be. I loved it down there. And it was a way to make a living." Fishtown today, Stallman reflected, "has grown so much. With the gift shops and everything, it's just completely changed. Completely." Yet fish boats and commercial fishermen have managed to survive. They still fish out of Fishtown.

Now, a new entity in the Fishtown story intends to keep it that way. In 2007, when the Fishtown Preservation Society took eight buildings and two fish tugs

The purse seine boat *Argo* harvesting whitefish in Grand Traverse Bay, c. 1980. PHOTO BY TOM KELLY

Fishtown, 2010. With its 2007 purchase, the Fishtown Preservation Society acquired eight buildings, two fishing boats, and related licenses and equipment. PHOTO BY LAURIE KAY SOMMERS

under its care, a new "season" of the fishery began. Its mission is to "preserve the historical integrity of Fishtown as a publicly accessible and authentic connection to local and regional history, Great Lakes commercial fishing, and maritime traditions and experiences." Or, to use the words of Marilyn Stallman, Fishtown Preservation strives to ensure that, for seasons to come, "It's still Fishtown."[19]

Chapter 5

Fishing Generations

"It takes more than one generation to make a fisherman."[1] – Bill Carlson, 1977

ROY BUCKLER, THE ELDEST SON OF FISHERMAN WILL BUCKLER, CAME OF AGE during Fishtown's heyday. Fishtown of the 1920s was a rough and tumble man's world, with new fish firms building shanties and ice houses and new boats taking shape in the boat yard. Gulls wheeled overhead, and the landscape was strewn with buoys, fish boxes, net reels, and (to the delight of gulls) fish guts. The sons of fishermen typically followed in their fathers' footsteps, but eighteen-year-old Roy dreamed of higher education. He spent a year at Central Michigan Normal School in Mt. Pleasant, but found that college wasn't for him. The fishing life, he realized, "was born into you." His father had been fishing out of Leland since 1908. In the fall of 1925 his father told him, "'I don't think you're ever going to make it in college,' and says, 'We need some help fishing, so you can go to work for me and see how you like it.'" Roy later reminisced, "That was the start of my fishing career. I was with him for about a year and a half, and then my younger brother George went to work for my dad. After three years he says, 'I think you fellows can run the fish rig, so I'll turn it over to you.' So from then on he used to go with us once in a while. When he thought we weren't doing the right things, he'd try to teach us more tricks of the trade. But that was about the last year he went on the water with us. It was ours from then on."[2] Buckler spent fifty years on the lake and later mentored his own son, Terry, who became his fishing partner

Mackinaw boat sails into Fishtown, c. 1900. COURTESY OF LEELANAU HISTORICAL SOCIETY

and the third generation of Bucklers to fish out of Fishtown.

Roy and George Buckler's apprenticeship with their father was a common pattern in fishtowns. Lake Michigan could be hard and unforgiving. To survive and be successful, each generation depended on knowledge inherited from its predecessors. Mentors might be relatives or other fishermen. During the 1970s Brian Price crewed with father and son team, Fred and Ross Lang, and realized that his college degree went only so far on a fish boat. "It's not a random 'go out on the lakes with your nets and hope for the best,'" Price emphasized. "These people have generations of lore that they're going by, plus their own observations and their own gut instincts about where they can find fish."[3]

What of those first generations of commercial fishermen? Who taught them the tricks of the trade? Irish-born Michael Daly, Leland's earliest known commercial fisherman, likely had previous experience as a saltwater sailor and fisherman, as was the case with the Irish of the nearby fishing centers of Beaver and Mackinac

Islands. Charles Allard, the second Leland fisherman to be listed in the *Michigan State Gazetteer*, was born on Mackinac Island and sailed the Great Lakes for fourteen years before turning to fishing. Other pioneering fishermen, who hailed from landlocked European homelands or American states, must have had a steep learning curve in the crucible of the Big Lake. They braved the roiling seas of Lake Michigan in wooden Mackinaw boats no longer than 28 feet. Although known as the best surf boat of its era, a Mackinaw's open deck left fishermen exposed to the elements. The Manitou Islands were ten to sixteen miles distant, across the Manitou Passage where the bones of many boats and schooners lay on the lake bottom. It took guts, skill, and luck to navigate the lake.[4]

During the nineteenth and early twentieth centuries, European American fishermen shared the lake with their American Indian counterparts. For generations, Ottawa bands fished for subsistence in Lake Michigan and the streams

Nineteenth-century fish camp at South Manitou Island, Lake Michigan, with a shanty, Mackinaw boats, and gill nets on the reel. More fishermen worked from the islands during this period than they did from the mainland at Leland. IMAGE COURTESY OF NOAA HISTORIC FISHERIES COLLECTION, NOAA NATIONAL MARINE FISHERIES SERVICE

of what is now Leelanau County from *jimanas* (small boats) and white cedar canoes covered with birch bark. (Ojibwa began settling the Grand Traverse region in the 1800s.) The graveled shallows of the Carp River, rimmed by swamp cedar, were fertile spawning grounds for Lake Michigan fish. The Ottawa band living at Carp River would have worked cooperatively, as was Indian custom, with clearly defined gender roles. The men made gill nets from fibers gathered by women; men did the actual fishing while women smoked or dried the catch on triangular racks. These Indian fishing skills and technologies shaped the way European Americans later utilized the Great Lakes for commercial fishing.[5]

Great Lakes Indians also engaged in commercial fishing, continuing a practice that began in the 1600s when Anishinabek peoples traded fish with the French. Indian craftsmen in Peshawbestown, located across the peninsula from Leland, built a Mackinaw boat for commercial fishing as early as 1850. Local Indians continued commercial fishing until around 1910, when they could no longer compete with their white counterparts. Early generations of Leland fishermen would tell stories about "fishing with the tribes," as remembered by resorter Dick Ristine, suggesting an unwritten history of interaction through the early 1900s.[6]

Leland's early commercial fishing fleet had only a handful of fishermen, its size curtailed by the lack of a natural harbor, industrial use of the river and lakefront, and strong prevailing winds off the Manitou Passage that limited the range of the old sailing boats. The river did not yet support a full-scale fishing village; the *Michigan State Gazetteer* did not even list fish as one of Leland's exports until 1897. By the 1880s, when Charles Allard joined Michael Daly in Leland, North Manitou Island had nine fishermen, Charlevoix 35, and Beaver Island—which in 1885 was the nation's largest freshwater fishing port—164 full-time fishermen, 23 fishermen/farmers, and 30 on-shore workers who cleaned, iced, and packed fish. Many of Leland's nineteenth-century fishermen were "beachcombers" who fished for subsistence or to supplement income. They used light gear, smaller boats, and worked off shore or from shanties along the coast. Even this could be treacherous. Swiss immigrant Minrod Buckler, a Civil War veteran and grandfather to Roy, drowned while fishing off the beach. Roy Buckler did not consider his grandfather a true commercial fisherman: "Everybody had a net that used to go down, and they set it on the beach to get some fish to eat. You didn't buy a fish; you'd go down, caught it yourself."[7]

The fishermen at Fishtown viewed themselves as a brotherhood, linked by occupation, geography, and in many cases by ethnicity and family. Like other settlers in northern Michigan, Fishtown's fishing families came from diverse homelands: Clausses and the Maleskis from Prussia; the Steffens, Hartings, and Lights with roots in Germany; the first generation of Bucklers from Switzerland; the Cooks with roots in France and Sweden; the first generation of Carlsons from Sweden; the Masons with roots in Scotland; the Firestones from Indiana; the Smiths with roots in Massachusetts; the Guthries with roots in New York state and Ireland; Brown and Kaapke with roots in Bohemia; Severt Johnson from Norway; the Prices from England via Canada; and Peter Nelson with roots in Denmark and Sweden.

Some—like the Carlsons, Firestones, Johnsons, Bucklers, and Maleskis—had known one another on the Manitous before moving to the mainland. Others formed partnerships with siblings, fathers, uncles, sons, or in-laws. Those related by marriage included Buckler and Steffens, Carlson and Firestone, Firestone and Kaapke, Harting and Light, and in later generations, Maleski and Price, and Price and Buckler. There were fathers and sons: Bucklers, Cooks, Steffens, Carlsons, Prices, Hartings, Clausses, and most recently, Langs. The Carlsons

Fishermen at the Fishtown reel yard, c. 1920. COURTESY OF LEELANAU HISTORICAL SOCIETY

have had five generations involved in commercial fishing; the Bucklers four. The Steffens, Maleskis, and Prices continued through three generations. In a small, close-knit community this continuity has been essential to Fishtown's survival. Some fishing partnerships lasted years, others were more fluid. As Bill Carlson remembered, "A lot of these unions changed, you know. People would fish with someone for a while, and then go fish with someone else. So I remember them well. These are all great, intelligent, strong people. There were quite a few alcoholics in fishing, and probably for good reason, but I remember those guys, too. They were real characters."[8]

The men shared equipment and occupational knowledge, and deeded shanties to one another for the modest fee of one dollar. Dick Ristine recalled the repartee in Fishtown during the 1920s and 1930s: "I would often be down on the dock, and they would talk back and forth across the river about where they caught fish that day. Of course they had no radar and no walkie-talkies or cell phones. It is remarkable to me how they communicated and seemed to share their experiences with each other, not in any competitive way but in a very co-operative way."[9] Regardless of relationship, fishermen had to have each other's back. They didn't always get along, but in times of crisis, they were a brotherhood. The occupation was too dangerous, the lake too quixotic for it to be any other way.

Harting Family and Otto Light

For the most part, the life experiences of the early fishing generations have been lost with time, unless, as in the case of John Harting (1855-1908), a horrific event thrust his name into the local newspapers. Harting (also spelled Hartung) had fished out of Leland since at least 1889. The Midwest-born son of Bavarian immigrants, Harting alternated commercial fishing with captaining the steamer *Tiger*, one of the Lake Leelanau boats that transported resorters, farm produce, freight, and mail between Fouch (a stop on the Manistee and Northwestern train route) and Leland. The *Tiger* started her runs on the lake in the spring of 1893, but by 1907 competition with the rival Lake Leelanau steamer, the *Leelanau*, drove the *Tiger* out of business. On 16 August 1908, Harting, then back to fishing, boarded the *Leelanau* for Traverse City to pick up parts for his fish boat, likely the *Morning Dip*, built in Leland for the Hartings in 1905. The *Leelanau's* captain, Charles Mosier, asked Harting to take the wheel while Mosier and his engineer

The *Ace* moored in front of Harting and Light's shanty, c. 1950. PHOTO BY JOAN FISHER WOODS

went below to make adjustments to the engine. "Suddenly the boiler 'let go with a roar like a cannon,' sending pieces of the pilothouse flying and covering the boat in a shroud of steam. The *Traverse Bay Eagle* reported, 'At first it was believed that Hartung had been killed instantly. Mr. Cook and others who saw him as he laid limp, unconscious, partially outside the pilot house, and bleeding terribly from many wounds, said it was a very sickening sight. He was partially blown through the window.'" John Harting died the following day.[10]

During the period of time in which the accident happened, Harting and his son, Will, (1882-1955?) were fishing partners. The younger Harting continued

Harting and Light's former shanty, currently known as the Crib retail shop, 2010. PHOTO BY LAURIE KAY SOMMERS

fishing with various new partners until 1926, when he joined forces with his brother-in-law, Otto Light (1895-1963). The two purchased Fishtown Lot 9 in 1926 and built the shanty today known as the Crib. The following year, they hired Omena boat builder John A. Johnson to build a new fish tug, the *Ace.*

Harting and Light evidently suited one another. They fished together for twenty-three years, perhaps the longest continuous partnership in the history of Fishtown, until advancing years and the lamprey incursion prompted their retirement. They were remembered as "solid guys" and somewhat taciturn. Summer resorter Dick Ristine, who crewed with Harting and Light on the *Ace* in the summer of 1938, recalled Otto's cryptic humor: "I remember [Otto] dipping a cup over the side into the water, drinking it and then tossing the water away, saying 'Mighty thin.' That was about all that I remember Otto ever said." Light, a bachelor until late in life, had been gassed in World War I while serving as part of the 32nd "Red Arrow" Division, a fierce fighting unit of Michigan and Wisconsin National Guard that served on five major fronts in three offensives. He returned from the war and became a fisherman. It is easy to imagine how the dangers of a fisherman's life were tempered by the experiences of war. "Otto had a little power boat called the *Red Arrow*," remembered Ristine. "He would take a drive by himself almost every evening in the little lake [Lake Leelanau]. That was his life. But there were a lot of people, who had been gassed in World War I, who had very restricted and sad lives. I expect his was as fulfilling as some."[11]

Price Family

The Prices were among the earliest fishing families in Leland, reportedly setting nets in the 1870s. During the late nineteenth century, the primary catch was whitefish. On its front page the *Leelanau Enterprise* noted exceptionally large lifts, like E. B. Price's 1,200 pound haul from a single net in 1897. Egbert E. "Dad" Price

E. B. Price boiling and mending gill nets behind his shanty, late 1800s. COURTESY OF LEELANAU HISTORICAL SOCIETY

(1841-1917) was a massive man. He stood 6 feet 9 inches tall and was described as "standing an axe handle wide at the shoulders." By 1875 he had established himself as a cooper in Leland, following a common pattern of combining fishing with another trade. Price's cooperage likely crafted barrels used to ship brined fish to markets up and down the lakes.[12]

Price built one of the earliest shanties at the river's mouth, from which he ventured out to the Lake Michigan fishing grounds in his sailboat, *Macknac*.[13] In 1900 he turned the reigns over to his sons, Warren (1874-1956) and Oscar (1877-1942), so the Price family, too, continued as part of the Fishtown story. The brothers eventually parted ways in 1915 and worked out of separate shanties and fish tugs.

Warren Price on dock, Joe Nedow in boat, c. 1905. COURTESY OF BARBARA GENTILE

Warren Price's shanty (far right), c. 1930. COURTESY OF LEELANAU HISTORICAL SOCIETY

The Warren Price Shanty was acquired by the Carlson family in the 1930s. Summer resorter Bill F. Hall purchased the shanty from the Carlsons in 1956 and it is today known as the Hall Shanty. PHOTO BY LAURIE KAY SOMMERS

Warren Price had married Clara Buckler in 1904, sister to fisherman, Will Buckler. About 1918, Warren replaced the old Price shanty (frequently damaged by storms) with a combination fish house and ice house. He fished with new partner, Will Carlson, interrupted by a six-year hiatus when he took summer excursion parties on Lake Leelanau as a fishing guide. At some point in Price's career, he briefly worked as a fish buyer. During the 1920s, along with the Hartings and Otto Light, he and Clara spent winters in Florida, where Warren had a real estate business.[14] He was forced to quit fishing in the mid-1930s due to a debilitating stroke. His brother, Oscar, continued the Price fishing tradition, working with sons Norman (1908-1998) and Vero (1913-1986).

Oscar Price was an amiable soul who had been fishing since age thirteen. Dick Ristine remembered him as the friendliest of the local fishermen, citing an incident when a group of men were bringing a big boat ("maybe 40 feet long") over the Fishtown dam. "They were using block and tackle to try to pull this thing," Ristine recalled, and "hooked the block to Oscar Price's net drying posts." The force "snapped them off, which didn't seem to bother Oscar; he just smiled!"[15]

In 1919 Price and then-partner Will Harting acquired a new boat, the *Wolverine*, which became Oscar's alone when Harting and Light started their fish firm. The boat served him well until a fateful day in the summer of 1934. On 7 June the

Oscar Price's ill-fated *Wolverine*, c. 1920, moored in front of Price's shanties. The two buildings once occupied the site of the Louis Steffens Shanty, known today as the Village Cheese Shanty. COURTESY OF LEELANAU HISTORICAL SOCIETY

Leelanau Enterprise carried the dramatic headline, "*Wolverine* burns, Price and Son Escape Death by Minutes, Steffens to the Rescue, Two Thousand dollar loss hits Leland fisherman at bad time." The newspaper described how the two had clung "to the edge of a burning boat or swam in the ice cold water until nearly exhausted. Mr. Price says that he noticed that the carburetor on the motor was leaking, and the men decided to leave the rest of the nets and start for home. As Price cranked the engine to start the boat there was a burst of flame and almost instantly the interior of the boat was afire."

Bruce Price heard the story of how Oscar told his son, "'You swim for it; you have two children.' But Vero replied, 'No, I'll stay here with you.'" Oscar suffered severe burns on his face and neck. Luckily, Henry Steffens and Will Wichern in the *Helen S* were just two miles away and glimpsed the first puff of smoke from the burning boat. Bruce Price recounted, "The *Helen S* made it over there and got them, and towed on to what was left of the bottom of the boat. Supposedly it sank just outside of Leland Harbor. That's the last story I had."[16]

By fall, Oscar Price launched his new boat, the *Nu Deal*, crafted in Fishtown's boat yard by Adolph Johnson, son of Omena boat builder, John A. Johnson. Two differing stories exist regarding funding for the *Nu Deal*. Oscar Price's obituary

Oscar Price, son Norman, and Otto Light checking a pound net. COURTESY OF LEELANAU HISTORICAL SOCIETY

in the *Leelanau Enterprise* states that the boat "was made possible through the kindness of the Prices' many friends who volunteered to aid him in replacing his loss." An alternate version comes from Bruce Price. "My grandmother had so much pride that she wouldn't take the gifts from Leland. She went in through the Farmers Home Administration, which was the New Deal under Roosevelt, and then she named the boat the *Nu Deal*. What was kind of surprising about that is the fact that the New Deal was a Democratic program, and my grandmother, Emma, she was a staunch Republican. So I'm surprised she did that. I don't think she really realized it at the time."[17]

Oscar Price eventually suffered a stroke as a result of the accident with the *Wolverine*. He died in 1942. Vero escaped physical harm but had lasting psychological scars and ultimately was institutionalized. Norman Price tried to continue, but fishing was poor. By the early 1950s the abandoned Price shanties were sliding toward the river. Norman sold the property to Louis Steffens in 1956, thus ending the Price family era in Fishtown.

Cook and Brown

"Captain Cook looked like an old salt," Dick Ristine remembered. "He lost part of his hand, always had a pipe in his mouth, a grizzled white beard." The missing fingers are part of Fishtown lore. The speculation is laid to rest by this succinct account from the *Leelanau Enterprise* 5 July 1900: "George Cook had his right hand blown off yesterday by the discharge of a large cannon cracker. Dr. Fralick of Maple City dressed the injured member." The fishermen had been part of the Independence Day parade, and the twenty-seven-year-old Cook literally got caught up in the festivities. Despite the disability, George pursued a successful career as a commercial fisherman. "He was amazing with what he could accomplish with a thumb and a finger or two," recalled resorter Barbara Gentile, whose family became personal friends of Cook.[18]

George Cook (1872-1965) was perhaps the most colorful of all Fishtown fishermen. In later years, he delighted in posing for tourist photos and artist sketches. His long career included most occupations common to northern Michigan. Before pursuing commercial fishing he worked as a dock walloper, lumberjack, carpenter, and farm worker. Cook was born in a log cabin on Lake Leelanau to French immigrant Frederick Cook and Aurora Lundgren, a native of Sweden.

George Cook at the net reel, while tourists watch, 1950s. COURTESY OF TRAVERSE AREA HISTORICAL SOCIETY

Aurora took the young George back to Sweden for an extended visit when he was young, but he then spent the rest of his life in Leland. By the time he married Christine Anderson of Centerville in 1904, he likely had been fishing for a decade, beginning his career with John Harting and Napoleon Paulus. A year before his marriage, Cook and new partner Martin Brown had built the shanty on the site of the present-day Otherside Vacation Rentals. (Portions of the original shanty remain in the interior framing.) Brown (1871-1964) first sailed the Great Lakes at age fourteen and then joined the Life Saving Service on North Manitou Island. During his partnership with Cook, which lasted until 1920, he also pursued a career in local politics and eventually left Leland for Grand Rapids to become U.S. Marshal for West Michigan.

The 1903 completion of Cook and Brown's spacious shanty—handsomely clothed in cedar shingle—marked the beginning of Fishtown's new era of expansion. George continued in the trade until the early 1940s and subsequently worked as a fish buyer and shipper until 1963. He passed his rig, gear, and fishing expertise to his son, Marvin (1914-1991), ensuring continuity into the next generation.

The Otherside Vacation Rentals, pictured here in 2010, incorporates remnants of the George Cook and Martin Brown Shanty. It was remodeled in 2001 to the plans of Suttons Bay architect David Hanawalt. PHOTO BY EVAN HALL, COURTESY OF HOPKINSBURNS DESIGN STUDIO, ANN ARBOR

Cook and Brown's new shanty (second from left along the river), pictured c. 1906, with Alex Mason's shanty to the far left, and the Price and Clauss shanties to the right. The building to the rear is an ice house. On the north side of the river, the Carlson and Johnson shanty sits closest to the lake, next to a shanty possibly built by the Prices and later occupied by brothers William and Lawrence Smith. COURTESY OF BARBARA GENTILE

Buckler Family

William Buckler (1873-1940) was born in Good Harbor. As a young man, he worked as a mail carrier and farm hand before moving to North Manitou Island to join the Life Saving Service. While on the island he worked building cottages, planting fruit orchards, and from 1900 to 1905 as lighthouse keeper. In 1903 he married Matilda Steffens, known as Tillie, who was the sister of Leland fisherman, Henry J. Steffens. Their eldest son Roy was born on the island in 1906. The family subsequently moved to Leland, and Will took up commercial fishing. His first partner, Will Cook (brother to George), had a penchant for the bottle. "As soon as he got off the boat," Roy recalled, "[Cook] was going to get a drink of whiskey up at the saloon. It didn't work very good. That lasted about a year and a half. My dad bought him out and went by himself." Subsequent (and more sober) partners included Henry Steffens and Oscar Price. During the late 1910s, an ear infection forced Buckler off the water for three years, but as soon as health permitted he returned to commercial fishing. He moved his

Will Buckler cut his shanty in half in 1928 and added a section to the middle, visible with the lighter colored cedar shake. His tug, *Irene*, is moored alongside, 1930s. PHOTO BY ERHARDT PETERS, COURTESY OF LEELANAU HISTORICAL SOCIETY

A night view of the former Buckler shanty, now owned by the Leelanau Historical Society, 2010.
PHOTO BY LAURIE KAY SOMMERS

Terry and Roy Buckler removing chubs from the smokehouse, 1960s. COURTESY OF LEELANAU HISTORICAL SOCIETY

operation across the river in 1921 after purchasing south-side Lots 2 and 3 from fisherman William Smith (1887-1947). Two shanties occupied the lots. During the late 1920s Buckler tore down one and cut the other in half, adding more space in the center. Sons Roy and George took over the family fishing business, and Will began a new venture with a filling station he built on the corner of River and Main Streets in Leland.[19]

Roy Buckler fished with Pete Carlson for three years during World War II while his brother George was in the service. When the fishing went bad in the late 1940s, Roy deeded the shanty to Marvin Cook and his partner John Maleski Jr. and left fishing for factory work in Traverse City. Fishing was in Buckler's blood, however, and by the late 1950s he had returned to Fishtown, this time with his son Terry (1940-2006) as partner. The Bucklers persevered through the lean times of the 1960s and 1970s, with Terry ultimately carrying on the Buckler tradition by working for his dad and providing fish for Steffens and Stallman. Terry Buckler also was part of the purse seine experiment of the late 1970s and early 1980s with fellow Leland fishermen Bill Carlson and Ross Lang. Terry finally left Fishtown to fish chubs and perch in southern Lake Michigan. His son Jason worked for the Carlsons and represents the fourth generation of Bucklers to fish in Leland.

Roy Buckler remained part of the Fishtown family into the 1980s. As remembered by Alan Priest, "Roy worked on our nets till he was over eighty years old. Then one winter he slipped and broke his hip, and his health went down from there." For Buckler, fishing had been a good life. "We always had something to eat," he once said. "If I had to make a choice again, I think it would be the same. I would be a commercial fisherman."[20]

Kaapke and Firestone

Claude Kaapke (1897-1967) formed a fishing partnership with the Firestones (father Eli [1873-1963] and son Roy [1901-?]) in the late fall of 1926. Together they built a new shanty on Lot 8 on the river's north bank (today the fish processing portion of Carlson's of Fishtown). Kaapke had married Eli's daughter, Hazel, in 1923. By 1926 he was clearly a man in search of a job. That year he had been employed in Flint by Buick Motors, moved to North Manitou Island to work as assistant light keeper, and by the end of 1926 had started a fish firm with the Firestones. Eli probably never fished; rather, he financed the operation for his son and son-in-law. Throughout the following decade, the *Leelanau Enterprise* mentions only Roy as a commercial fisherman.

The Firestones had come to North Manitou Island from Goshen, Indiana, in 1894. Eli Firestone married Emma Carlson, daughter of Nels and Sophia, on North Manitou Island in 1898, creating the initial connection between the Firestone and Carlson families. Another family tie occurred when fisherman Will Carlson married Anna Nerland; Eli Firestone's brother, Albert, was married to Anna's sister, Marie. In the summer of 1921, Nels Carlson sold his "fine large farm" outside Leland to Eli and Emma, and Eli likely spent the rest of his career in agriculture.[21]

Since Kaapke and the Firestones had not been born into fishing, Emma's brothers, fishermen Will and Edwin Carlson, may have helped them learn the ropes. By 1930 Claude and Roy had learned well and were averaging 750 trips per year in their new Leland-built tug, the *Etta*. In February 1931, they nearly met their end in the boat, when the *Etta*'s engine failed due to a damaged carburetor, and her crew was cast adrift in Lake Michigan's frigid waters for ten hours.

The *Etta*, moored by the Kaapke and Firestone shanty (far right), as Henry Steffens and the *Helen S* head out to the fishing grounds. c. 1930. Today this shanty is the dockside portion of Carlson's fishery. PHOTO BY ERHARDT PETERS, COURTESY OF LEELANAU HISTORICAL SOCIETY

"Kaapke said that at dark they were in the channel of the big boats," reported the *Leelanau Enterprise*. The men used their clothing and oilskins for flares, lighting seven in all, but "intermittent snowstorms prevented the watchers at Leland from seeing any of the signals until late in the evening." A worried Eli Firestone notified the Coast Guard to start a search, but it was Henry Steffens in his boat, the *Helen S*, "with a crew of volunteers that twice set out from Fishtown into the stormy lake in the direction of the last flare." The *Enterprise* continued, "A spotlight had been put at the harbor to show Steffens the way in through the storm. The mariners were none the worse for their adventure, except that they were hungry and none too warm."[22] Perhaps such adventures ultimately took their toll. Kaapke left fishing in 1944, likely to serve in World War II, and he never returned. The obituaries for both Eli and Claude list them as retired fruit farmers with no mention of commercial fishing. They deeded their shanty to Emma Firestone's brother, fisherman Edwin Carlson.

Maleski Family and Peter Nelson

The *Leelanau Enterprise* for 25 October 1928 featured the headline, "Leland Now Has Eight Fish Rigs, John Maleski and Peter Nelson Build Shanty on North Side, Will Fish this Fall." The paper reported that "Peter Nelson, living south of Lake Leelanau, and John Maleski, who recently moved here from North Manitou Island, have formed a partnership. Nelson and Maleski have built a shanty on the north side of the harbor, west of the Firestone and Kaapke place of business. This makes a total of five shanties on the north side of the harbor, while two years ago there were only two. The new partners have purchased a boat in Northport, formerly used by Edwin Carlson. Both have had experience in fishing."

John Maleski (1885-1950) was born and raised on North Manitou Island, where he had a farm and fish camp on the island's northeast shore. For several years before moving to the mainland, he worked for the Manitou Island Association (MIA), the Chicago-based syndicate that owned much of the island. Selling his farm to the MIA and relocating to Leland caused a rift with his brother Paul because John didn't keep the land in the family.[23]

Peter Nelson (1883-?) was raised in Good Harbor and initially crewed for a local fisherman out of Good Harbor Bay. Nelson later worked in lighthouse repair and as an itinerant carpenter. According to his son, Herb Nelson, Peter's parents "wanted him to come home and farm, but it was such a poor farm he couldn't scratch out a living." Nelson purchased another farm but returned to carpentry and commercial fishing as his livelihood. "He loved the water, he loved to fish," Herb recalled. His first Fishtown partnership was with Fred Anderson. After that dissolved, Nelson in 1926 purchased Lot 7 in Fishtown next to Kaapke and Firestone. When John Maleski arrived in Leland, the two became partners and built their new shanty near the river's mouth. By the fall of 1929 John wanted to buy out Nelson's half of the business. Herb recalled how Maleski and his daughter Gertrude trudged several miles through the snow "across Lake Leelanau and came to our house to pay my dad off for his share of the fishing business." Maleski prudently had waited until after the first of January, so he could draw a full year's interest before making his payment. Peter Nelson returned to carpentry.[24]

Maleski fished until his death in 1950. He preferred to work alone and stay closer to shore. Dick Carlson recalled that, in lieu of a crew, "he used to have kind of a contraption that would help him set the nets. He'd run the corks on top of this rim, the floats would go down, and this little mechanical thing worked like another person with him."[25] Maleski's son, John Jr. (1919-1975), later crewed on

Nelson and Maleski shanty, c. 1940, after John Maleski had assumed full ownership, with his tug, the *Bonnie Lass*, moored alongside. COURTESY OF LEELANAU HISTORICAL SOCIETY

The relocated Peter Nelson and John Maleski Shanty, pictured here in 2010, now serves as the retail portion of Carlson's of Fishtown. PHOTO BY LAURIE KAY SOMMERS

Fishtown tugs. Maleski's nephew, Jeff Houdek (1951-2010), worked for Carlsons and became captain of the *Janice Sue*.

Steffens Family

Henry J. Steffens (1888-1976) grew up on a farm in East Leland and in 1910 began fishing for brother-in-law William Buckler. Steffens joined the Coast Guard in 1913 but missed the freedom of commercial fishing; he returned home and rejoined the fishing fleet. Steffens acquired the deed to the Johnson and Carlson shanty in 1919 from the retiring Nels Carlson. He and hired man Bill Wichern became Leland's first "film stars" when they appeared in a 1926 Reo News-Reel film "showing the pond [sic] net fishing and the fishermen in action. Late in July a photographer, representing the Reo Motor Company of Lansing, came to Leland for the purpose of getting motion pictures of the fishing industry."[26]

Steffens' daughter, Marilyn Steffens Stallman, remembered her father as a strong, quiet man. "He wasn't afraid of anything. He had a good head on him." In 1927 Steffens invested in the *Helen S*, the 35-foot tug built by John A. Johnson that would see him through to his retirement in 1954. Steffens' "good head" enabled him to survive the "worst blow he remembered," the big Armistice Day storm of the late 1930s, when he lost ten boxes of nets valued at one hundred dollars each. At the helm of the *Helen S*, Steffens rescued the crew of the disabled *Etta,*

Henry J. Steffens roping nets, c. 1930. PHOTO BY ERHARDT PETERS, COURTESY OF LEELANAU HISTORICAL SOCIETY

The Steffens shanty with the *Helen S* moored alongside, c. 1940. Henry J. Steffens acquired the building from Nels Carlson, who built it in 1905 with his fishing partner, Severt Johnson. COURTESY OF LEELANAU HISTORICAL SOCIETY

and pulled the Prices from the lake after the *Wolverine* burned to the water line. After retirement from commercial fishing, he pursued a business smoking chubs. "He'd peddle the fish all over the county in their 1949 Studebaker," Bill Carlson recalled. "He might sell forty or fifty pounds out on his peddling route."[27]

Steffens' two sons, Henry Jr. (Hank) (1927-1988) and Louis (1921-1964), continued in the fishing trade. In 1958 they introduced the first steel-hulled tugs to Fishtown, the *Mary Ann* and the *Janice Sue*. Louis Steffens was a bachelor who lived with his parents and took care of them until his untimely death in 1964 from complications of lung cancer. He fished out of the old Cook and Brown shanty, while letting his father smoke fish in Louis' new shanty, built about 1959 on the site of the former Price shanties. Artist David Grath remembered Louis as "a man of very steadfast habits" whose daily ritual was to visit the Bluebird Restaurant bar after he came in off the lake. The girls working at the bar would have a cigar waiting in the ashtray and a freshly opened Carlings Black Label beer. "Louis would come in and just nod quietly," Grath recalled. "Very rarely said much about anything, but he loved the attention. Who wouldn't?"[28]

As recalled by Bill Carlson, Louis' brother Hank "was a well-educated, second-

The Henry J. Steffens Shanty after its 2010 rehabilitation to the plans of architect Richard A. Neumann of Petoskey. The shanty currently houses the Diversions retail shop. PHOTO BY LAURIE KAY SOMMERS

Louis Steffens' shanty, pictured at right, mid-1960s, was the first new shanty in Fishtown since the late 1920s. He died before using it. PHOTO BY MIKE BROWN

generation fisherman. He went to Michigan State and got a teaching degree. They fished in the summertime, and then he was going to teach in the off-season. His partner was Leo Stallman [1923-1979], which was his brother-in-law, and that's how that partnership came about. Real characters." Hank was as social as his brother Louis was quiet. During the 1950s he met Michigan State University Summer Art School student Ann Lyman at a party. Ann, then engaged to someone else, was swept off her feet by the handsome young fisherman, and they were married the following St. Patrick's Day. During the botulism scare of 1963 Ann and Hank went back to graduate school, and Hank worked as a teacher in Marlette, Michigan. His active fishing partnership with Leo Stallman Sr. finally came to an end in 1968, when the Department of Natural Resource's new retroactive limited entry law forced part-time fishermen like Steffens and Stallman out of business. Hank and Ann moved to Clearwater, Florida, where he worked as a school teacher. Leo Stallman and his son, Leo Jr. continued to sell smoked fish

Louis (left) and Hank Steffens, 1945. COURTESY OF GLENN GARTHE

The former Louis Steffens Shanty, currently the Village Cheese Shanty, 2011. PHOTO BY LAURIE KAY SOMMERS

out of their shanty until 1973. Steffens and Stallman subsequently rented their former fishing buildings for retail use, ultimately selling them in 1977 to Carlson Properties.[29]

Carlson Family

By virtue of longevity and survival, the Carlsons are synonymous with Fishtown, tracing their roots to the early 1900s, when Swedish immigrant Nels Carlson (1854-1935) became the first Carlson to fish out of Leland. Nels met his wife Sophia in Leelanau County and the two settled first on South Manitou Island and then on North Manitou, where they homesteaded a 160-acre farm. Nels did a variety of jobs to make ends meet: fishing (his listed occupation in the 1880 census), stock raising, subsistence farming, running the mail boat, serving as a volunteer for the Life Saving Station crew, and building a school in his spare time. While on the island, the Carlsons had twelve children.

As transportation to distant markets became less frequent, and large land owners gradually gained a monopoly on the island economy, Nels contemplated leaving for the mainland. His son, Edwin Carlson recalled how, on St. Patrick's Day 1904, "following an exceptionally severe winter, the Carlsons packed their possessions on three sturdy logging sleds and hitched teams of horses for the journey across the frozen lake." During the eight hour trip the family "kept the sleds widely separated in case of a breakthrough and ventured out across miles of ice with all the pigs, sheep, poultry and milch cows bringing up the rear." Nels' great-grandson, Mark Carlson, heard the story that each sled had a male family member to help in case of an emergency. When Nels and family woke up the next morning after arriving safely in Leland, the ice had completely vanished.[30]

Nels Carlson purchased a farm on the west shore of Carp Lake (now Lake Leelanau), but fishing proved his true calling. He worked with several early Leland fishermen until settling on partner Severt Johnson (1868-1949), who had served on the North Manitou Life Saving Station crew with Nels' eldest son, Will. Johnson, a native of Norway, had fished pound nets with George and Will Cook as early as 1900. In the late fall, 1905, he and Nels teamed up to build a shanty and ice

The original Carlson and Johnson shanty, located closest to the lake on the river's north bank, c. 1905.
COURTESY OF LEELANAU HISTORICAL SOCIETY

house. The two remained partners until 1914, when Johnson left for Gloucester, Massachusetts, to become a saltwater fisherman.[31] By 1919 Nels had transferred his shanty to the young Henry Steffens and his fish tug to his son, Will Carlson.

William "Will" Carlson (1878-1941) had married Norwegian immigrant Anna Nerland whose family also homesteaded on North Manitou Island. The young couple moved to Leland, where Will began fishing with the tug *Leland* until it burned and sank off Fishtown. Carlson's new boat, the *Diamond*, served him until 5 August 1941, when another fire had more tragic results, resulting in Will's death and loss of the boat. During the intervening twenty-five years, brothers Gordon (1900-1971) and Will fished together, followed by Will and son Lester William, known as Pete (1910-1988). (Another Carlson brother, Edwin, spent much of his fishing career in Northport.)

As Pete Carlson described it, "I was practically brought up in the fishing trade." Beginning at age sixteen, he fished during the summertime with his father "and a fellow by the name of Warry Price who he was in partnership with at that time. I stayed fishing until I was nineteen. I took three years into the Coast Guard, and that wasn't for me either. It was nice being near the water, but I'd rather go fishing."[32]

The Carlsons made it through the difficult 1940s and 1950s in part by catching and smoking the abundant stock of chubs, and by fishing part-time in southern Lake Michigan off Saugatuck. In 1944 they moved their base of operations to the shanty originally built and used by Kaapke and Firestone. By the mid-1960s, when the Carlsons acquired the former Maleski shanty to house their expanding retail trade, Gordon and Pete were fishing together. They moved the Maleski shanty behind their own fishery building and reconfigured the two shanties into one. After Gordon Carlson became ill in 1968, Pete's son, Bill, came home from college to help with the family fishing business.

Lester William Carlson II (born 1943), known as Bill, grew up around fishing. His earliest memories were when "they'd sit me in the corner, and I'd fill mending needles until my fingers were too sore to do it anymore. But I'd make myself fifty or sixty cents in a good afternoon." Father and son fished together until Pete retired to become Leland Harbor Master, and the mantle passed to Bill. "When fishing was really tough in the late 1960s and 1970s," Bill recalled, "it took a while to try to figure out what to do. We felt our future was in retail and local wholesale. In order to get the greatest economic return, I felt that we had to catch the fish, process them, and move them right to the consumer. We opened a store in Traverse City [initially run by Bill's younger brother Mark] which was the closest metropolitan area. We developed our tourism business here a lot more. We bought what property in Fishtown was available, and there was some available because of the fishermen being put out of business. That's how we were able to stay in business and actually thrive in the business."[33] While their new business model helped to reinvent the Carlson fishery as a sustainable enterprise, the purchase of Fishtown property created an alternative to the tear down and redevelop approach that threatened other fishtowns.

Beginning in 1977, Bill Carlson, with his brothers Mark and Leon as silent partners, bought former Fishtown shanties and built others in the style of the historic fishing village of the 1930s and 1940s. Bill organized the Fishtown Preservation Society in 2001 to promote and preserve the fishing heritage of Fishtown. The organization went through several iterations and in 2007 acquired the Carlson Fishtown properties.

Carlson's is Fishtown's sole surviving commercial fishing enterprise. This

Bill Carlson (left) and his father Pete in their shanty, 1973. PHOTO BY TOM FOX, COURTESY OF BILL CARLSON

Will Carlson (rear) and son Pete bringing the *Diamond* into port, 1930s. PHOTO BY ERHARDT PETERS, COURTESY OF LEELANAU HISTORICAL SOCIETY

family business nearly came to an end seventy years ago. For Will and Pete Carlson, 5 August 1941, probably started like any other day: meeting at the Fishtown docks before dawn and readying their gill-netter, the *Diamond,* for the long trip out to 16 Fathom Shoal, one of the traditional fishing grounds furthest from port. But this was not just any day. A gas line broke, and their boat was engulfed in flames. For Will Carlson, it must have been a moment of déjà vu. His first tug, the *Leland,* burned and sank off Fishtown in 1916. Then there had been the *Wolverine*, burned to her railings with father and son Oscar and Vero Price plucked from the lake, shaken but alive. But this time, rescue didn't come in time for Will Carlson.

"We had one fire extinguisher," Pete recalled, "[but] it doesn't do any good to put water on gas. So there wasn't anything we could do to put it out. We decided that we'd have to take to the water, my father and I, which we did, with two life preservers. The boat drifted away from us, and we started out for North [Manitou] Island." For more than eight hours they swam toward the island, some eight miles distant, keeping up their spirits by talking about the future and planning their new boat. "About noon or one o'clock, I don't know exactly when, my father died from exposure. And I kept him with me." Carlson carried his father in his arms for seven more hours. When the South Manitou Coast Guard boat went by, and Carlson couldn't attract attention, "I had to let go of my father then. In the meantime I was trying to make shore and, of course, knowing this afterwards, most everybody was out looking for us."[34]

Roy Buckler also had been near 16 Fathom Shoal that day. "Pete and his dad caught fire there about fifteen minutes from where we were, but we never saw anything. If we were going the other way, we would have seen it. We couldn't figure out why Pete didn't come [back]. We didn't use radios until the 1950s. You just kept track of everybody."[35]

Will Carlson, pictured here in 1934, drowned 5 August 1941, after his fish tug, the *Diamond*, burned and sank.
COURTESY OF LEELANAU HISTORICAL SOCIETY

The *Good Will* moored in front of Carlson's shanty, 1958. PHOTO BY BERKLEY DUCK III

Pete's wife, Rita, became concerned when the *Diamond* failed to return and called the Coast Guard station in Traverse City for a search plane. The man who could authorize the plane wasn't there. "So the fish boats were out looking for him," Rita recalled. "Everybody took some liquor with them, because they thought if they were in the water, his dad and my husband Lester [Pete], they'd have something to give them. About four o'clock in the morning, I think it was, that Marvie Cook and Percy Guthrie were running him down. They had drunk the liquor, and they were going right towards him. The moon was up, and Marvie says, 'See that path of light there? Let's follow that.' And they saw him in that path of light from the moon."[36]

Pete recalled, "Why, they just about ran me down! And it scared the pants off them, too, because they didn't expect to hear anybody, and I—just as they went by—I hollered at them. They found me then, which was twenty hours later. The Coast Guard boat found my father at eleven o'clock that next morning, so at least we found him."[37]

After his ordeal Pete Carlson didn't know if he wanted to continue fishing. "He was sick for a long time over that," Rita remembered. "He had carbuncles on his arms and hands. He was waking up at night, and he was always swimming. He thought he was going to make it to the island, but the current carried him out, because that's what he did do. He had nightmares."[38]

It was a pivotal moment in Fishtown's history. In spite of all the close calls on the lake, the fishermen had always soldiered on. Never before had one of their own drowned while fishing. At the tail end of the Depression, money for a new boat was tight. But when Pete learned that the community was raising money to buy him another boat, he mustered the will to make a trial run to see if he could handle being back on the lake. "He went out with Roy Buckler," Rita remembered. "He went out a couple times, and he settled down, so he figured he could do it. And that's how he went back to fishing with Gordon Carlson [his uncle], and Louis Steffens." The new boat was built in Omena by John A. Johnson. "We called him Big John," Rita recalled. "He was a big guy. And it was launched in Omena harbor there." Rita chose the name *Good Will* to honor both Will Carlson and "the good will of the people that got it for us."[39]

Carlson's of Fishtown, 2010. PHOTO BY LAURIE KAY SOMMERS

In a community that has known its share of heartache and close calls, this particular story has become Fishtown's formative narrative. It is the story of a son's terrible choice and his incredible strength of spirit and body. "My dad holds the lake's survival time record," Bill Carlson once told the *Detroit News*, "and if it weren't for his rugged constitution I wouldn't be here today." It is also the story of a community's resolve to buy Pete Carlson a new boat and invest in Fishtown's future. Of all the families of Fishtown, it is the Carlsons, who—through grit, luck, ingenuity, and a community's generosity—continued Fishtown's proud legacy of commercial fishing.

Chapter 6

Fish Boats of Wood and Steel

"The fish tugs are just amazing vessels. The good ones, anyway. The bad ones don't last."[1]
– Brian Price, 2008

As a young man in the late 1950s, artist David Grath camped on the beach at Fishtown and was awakened before dawn by the "characteristic chuffing sound" of the *Janice Sue*'s diesel engine. "These were large engines with huge flywheels," remembered Grath. "They turned over quite slowly and developed a lot of power without turning at a high speed, so they had a beautiful sound to them."[2] The chug of the *Janice Sue*'s engine is still the music of Fishtown. This iconic gill-netter, christened in 1958, is the older of the two active fish tugs at Fishtown. Along with the trap net boat, *Joy*, it continues the legacy of commercial fishing vessels based in the Leland River. Fishtown Preservation's first major preservation effort was to restore the *Janice Sue* and *Joy* so that commercial fishing could continue at Fishtown. Boats have been benchmarks in the fishery's development: from wood to steel, from sail to diesel, from as many as eight in the peak fishing period to two today. Over the years, the *Leelanau Enterprise* and other newspapers gauged the vitality of Fishtown by the number and ownership of fish rigs lined up in the river. Visitors to Fishtown still gravitate toward the river, docks, and boats.

The boats of Fishtown have been vital links between shore and lake work. In his book *A Good Boat Speaks for Itself*, folklorist Tim Cochrane described how "boats were humane tools for fishermen. They were work partners; highly esteemed, sometimes loathed, but always talked about." These vernacular boats used traditional designs and shapes crafted to meet the particular challenges of working on the Great Lakes. Specific boat types developed for different types of nets: pound, gill, and trap.[3]

The fishermen usually built pound net boats and pile drivers at Fishtown. Like much of the operation of the fishery, use of pound nets required cooperation. During the peak fishing period, fishermen shared two pile drivers, or stake-boats, used to drive or pull the stakes that secured the net to the lake bottom. Roy Buckler remembered that "each crew had to have [a pile driver], because everyone wanted it at the same time. So we used to go over into somebody's swamp and get a bunch of cedar logs and spike her together." Built in the form of a raft or flat-bottomed scow, the pile drivers had an A-shaped framework with a weighted tackle operated by hand or motor that was used to drive the stakes. Fishermen

The pile driver (left foreground) was used to set and remove the pound net stakes, c. 1930. The narrow pound net dinghies or skiffs (center foreground) were used to enter the "crib" or "pound" section of the net that held the trapped fish. COURTESY OF LEELANAU HISTORICAL SOCIETY

maneuvered a small skiff inside the stakes to lift their nets. Fishtown's pound net fishery continued from the nineteenth century until the 1940s when overfishing and the sea lamprey combined to decimate the whitefish population.[4]

Most of Fishtown's boats served the gill net fishery. The earliest gill net boats were sail-driven wooden Mackinaw boats, an open boat design native to the Upper Great Lakes. During the first decade of the twentieth century, Leland's fishermen followed the larger Great Lakes pattern of powering their boats first with steam and then with gas engines.

The early gas boats were small, like George Cook's *Why Not*, built in 1905 and

The *Why Not*, c. 1906, a transitional gas boat that added a gas engine to the Mackinaw boat design. By the end of the decade most fishermen had added enclosed cabins. COURTESY OF BARBARA GENTILE

"Big John" Johnson working in the Fishtown boat yard, 1930s. COURTESY OF LEELANAU HISTORICAL SOCIETY

measuring 26 x 8 feet. Fisherman Roy Buckler explained, "One of the reasons for the small boats was that if the boat drew over 3 feet of water, you couldn't get in the harbor." Two piers jutted out from the river's mouth, leaving a narrow 20-foot span at the entrance that necessitated a smaller boat.[5]

Fish boats such as these "were purchased and used by people who often knew the builder well and talked with him about what they wanted built," explained Tim Cochrane. "Vernacular boats were the products of local craftsmen meeting environmental and functional requirements with nearby materials, traditional aesthetics, and knowledge."[6]

Two Scandinavian-born boat builders adapted skills from the Old Country to the Leland fleet. Christian Telgard (also spelled Telgaard), a native of the Norwegian village of Stangvik, came to the United States in 1887 at age twenty-seven after an early career in sailing. He spent the first four years at the Life Saving Station on North Manitou Island. By 1896 he left the island for Northport, where he and his sons established a successful boat building business. (Son Martin left the trade and with his wife, Leone Carlson, founded Leland's Bluebird Restaurant in 1927.) John Mitchell, author of *Wood Boats of Leelanau*, noted, "Telgard distinguished himself as a prolific, first-tier craftsman" who modified Mackinaw boats to create some of Leelanau's first gas boats. A complete list of Telgard's boats for Leland does not yet exist. Those known include the following: "Mackinaw boat" (Warren Price, 1896); *Manitou* (Nels Carlson, 1897); "launch" (a term sometimes used to describe fish boats, Alexander Mason, 1905); and a cabin for the mail boat, *Lawrence* (John Paetschow, 1909).[7]

Telgard's competitor and sometime collaborator was Swedish-born John A. Johnson. Like Telgard, Johnson left North Manitou Island to take up boat building on the mainland and set up shop in the village of Omena. Unlike Telgard, Johnson's Fishtown boats are well documented by newspaper accounts, the photography of Erhardt Peters, and oral narratives. Known affectionately as "Big John" or "Bigfoot" for his height and shoe size, Johnson became a fixture in Leland, working as a carpenter in town, routinely repairing boats, and building most of Fishtown's wooden fish tugs as well as mail boats for Tracy Grosvenor. Fisherman Roy Buckler remembered Johnson as "an artist with the adze." Photographs of the Fishtown boat yard show the steam boiler with which he steamed and shaped the ribs out of oak. Johnson hewed the keels by hand out of a big hand-picked red beech or birch and used cypress for planking.[8]

Johnson's known Leland boats (some built by or with his son Adolph) include

Late 1930s postcard from the heyday of Leland's wooden fish tugs. Fishermen-built log jetties extended from the river's mouth, inside the federal government's new piers that were completed in 1937. Pictured on the south river bank left to right: Will Buckler's *Irene* (1926), George Cook's *Sambo* (1926), and Will and Pete Carlson's *Diamond* (1916?). Pictured on the north river bank right to left: Oscar Price's *Nu Deal* (1934), Henry Steffens' *Helen S* (1927), Will Harting and Otto Light's *Ace* (1927), Claude Kaapke and Eli Firestone's *Etta* (1929), and John Maleski's *Bonnie Lass* (1935). COURTESY OF TRAVERSE AREA HISTORICAL SOCIETY

the following: *Bob* (Tracy Grosvenor, 1919); *Sambo* (George Cook, 1926); *Irene* (Will Buckler, 1927); *Ace* (Harting and Light, 1926); *Helen S* (Henry Steffens, 1927); *Nu Deal* (built by son Adolph Johnson for Oscar Price, 1934); *Bonnie Lass* (built by Adolph Johnson for John Maleski, 1935); *Manitou* (Tracy Grosvenor, 1936); *Smiling Thru* (built for L. J. Strayer of Northport, 1936, and then remodeled by Tracy Grosvenor as his mail boat); *John A* (Tracy Grosvenor, 1930s); and *Good Will* (Pete Carlson, built in Omena rather than Fishtown, 1942).

These wooden boats, once the pride of the Leland fleet, gradually outlived their usefulness. The resting place of most of Fishtown's wooden fish tugs is a mystery. The whereabouts of a scant few are known. The *Ace* is reportedly motoring the waters of Lake Charlevoix. The *Etta* was placed in front of a Frankfort, Michigan, restaurant until she rotted away. The *Smiling Thru* was dropped from marine registry records in 1972, listed as "abandoned."

The bones of Oscar Price's *Nu Deal* lie somewhere in the Gulf of Mexico. The boat had been launched with great fanfare in 1934. The *Leelanau Enterprise* described the christening with "a bottle of amber-colored liquid upon the prow. In fact the launching ceremony was held up momentarily while Vero, son of Capt. Oscar Price, and first mate on the new vessel, went to a near-by grocery for the purpose of getting said bottle of amber-colored fluid." After serving the

The Fishtown boat yard, with Johnson's steam boiler and the hull of Oscar Price's *Nu Deal*, 1934. PHOTO BY ERHARDT PETERS, COURTESY OF LEELANAU HISTORICAL SOCIETY

The *Nu Deal* near completion, 1934. PHOTO BY ERHARDT PETERS, COURTESY OF BLUEBIRD RESTAURANT

Price family well, the boat was pulled into the boat yard, where it sat until the 1950s. "Then a gentleman from Suttons Bay come along, fixed it up, and took it to Florida for a fishing boat," recalled Bruce Price. "The last we heard, it sunk out in the Gulf of Mexico."[9]

The *Good Will*, crafted and launched in Omena in 1942, was the last wooden tug built for Fishtown. The Carlsons replaced her in the mid-1960s with the steel-hulled *Janice Sue*, and eventually sold her to a buyer who intended to run the *Good Will* as a cruiser. Today, her rotting timbers lie on the bottom of Betsie Bay, between Frankfort and Elberta.[10]

The story of the *Helen S* has a happier ending. The Steffens family fished with her through the late 1950s, when her aging hull was critically damaged by ice floes off Leland Harbor. Her near fatal mishap spurred Fishtown's transition to the era of steel-hulled tugs. After conversion to a yacht when her fishing days were over, the *Helen S* has come full circle and is now owned by Fishtown Preservation,

The *Helen S*, 1940s. The vessel was named for one of Henry J. Steffens' daughters. COURTESY OF LEELANAU HISTORICAL SOCIETY

which intends to restore her for interpretive display.

Steel-hulled fish tugs, like the one that replaced the *Helen S,* plied the Great Lakes by 1930 but were not widely used until after the Depression. They represented the next generation of fish boats, a continuum that began with the age of sail, transitioned through steam to gas, and ended with diesel. More reliable and powerful engines allowed the boats to reach distant fishing grounds. Mechanical lifters, widely adopted by the early 1900s, enabled crews to use more nets at greater depths. Boat designs also changed over time to accommodate larger numbers of gill nets and to provide protection against inclement weather; enclosed cabins, for example, enabled fishermen to process fish en route to port. Earlier generations of gill-netters fished primarily lake trout and whitefish—the high value species—but overfishing and the invasive sea lamprey decimated these populations. By using larger steel tugs, commercial fishermen could focus on bigger hauls of remaining species such as chub and perch.[11]

By 1958 the Leland fishing fleet had been cut to just four rigs, half the number of its peak. Two of the four had barely survived a "Night of Horror" the previous winter. The *Leelanau Enterprise* chronicled how a fast-moving ice mass clogged the approach to the harbor, smashing and imprisoning three wooden fish tugs: the *Good Will* with fishermen Gordon and Pete Carlson and mail boat captain George Grosvenor; the *Etta* with Louis Steffens and Marvin Cook; and the *Helen S* with Hank Steffens Jr., his partner Leo Stallman Sr., and teenaged crew member, Terry Buckler. The *Good Will* managed to "heave and buck and plow her way through the ice pack," finally making it to the Fishtown docks, where an anxious crowd awaited news of the vessels. The *Etta* tried to follow the *Good Will*'s path through the floe, but her engine wasn't as strong, and she became stuck in the grinding mass. "The *Helen S*, moving into the ice mass behind the *Etta*, received a near mortal blow as the pressure of the ice caved in a rib. She began to take water as her planking started to heave." Terrified by the roaring, grinding sound of the ice, seventeen-year-old Terry Buckler "thought we were all going to die."

The *Helen S* tried to tow the *Etta* free, but finally had to make a run in the open water back to the island or risk sinking in the frigid lake. At dawn, George Grosvenor and the *Smiling Thru* "smashed and battered his way out of Leland Harbor against the ice," hoping to free the *Etta*, but "both boats were imprisoned then, side by side." Finally, the Coast Guard ice breaker arrived, and with her steel hull, plowed through the ice mass, so the *Smiling Thru* could tow the badly damaged *Etta* into the harbor. "'It was the worst experience I've ever had,'" Louis Steffens of the *Etta* told the *Leelanau Enterprise*. "'I've been in trouble before but nothing like this.'"[12]

After their ordeal, Louis Steffens and partners Hank Steffens and Leo Stallman Sr. hired Jim Derusha of the Marinette Marine Corporation in Wisconsin to build two new boats with safer, sturdier hulls made of quarter-inch steel. The *Janice Sue* and the *Mary Ann* were the first steel-hulled boats to join the Fishtown fleet. The *Fishing Gazette* of the time reported, "The steel hulls on the new boats are the first full skeg design to be used in this region. As a result of this design the owners report they can work closer to their nets for a longer period of time without the amount of rolling and drifting they had experienced in the older, conventional design."[13]

Brian Price, who fished on steel tugs out of Leland, remembered how much

Fishtown, 1960, at the transition from wood to steel, with the wood-hulled *Seabird* and *Good Will* and the steel-hulled *Janice Sue* and *Mary Ann*. COURTESY OF TRAVERSE AREA HISTORICAL SOCIETY

confidence the fishermen had in these "iron boats." "You could roll those suckers just about over and they'd come back up. They were built for any kind of weather: they were built to work in ice, they were built to work at the nets in heavy seas, and to take a sea broadside. They had a lot of weight welded into the keel."[14]

The *Janice Sue* and *Mary Ann* were not only the first steel-hulled fish tugs in Fishtown, but they were also the first new boats since Pete Carlson's *Good Will* in 1942. Their arrival created quite a stir. The *Fishing Gazette* observed, "There are a couple of new girls at Fishtown in Leland, Michigan, but don't get the wrong idea just because the boys have been giving them the eye. The *Janice Sue* and the *Mary Ann* are behaving like perfect ladies, according to their owners." The fishermen built the cabins themselves and installed state-of-the-art equipment, electronics, and power. This included AM band marine radios at the insistence of local resident Ray Schaub Sr. Glenn Garthe recalled how, after the wreck of the *Carl D. Bradley* off nearby Gull Island in November, 1958, Schaub had "pushed and pushed all of the fishermen in Leland to buy a radio. He became a distributor, so to speak." Schaub had reminded Leland's fishermen that the *Bradley*'s mayday call came via its radio, and that similar calls might have prevented the sinking of the *Wolverine* in 1934 and the *Diamond* in 1941.[15]

The boat names continued the old wooden boat tradition of naming for female family members. The *Mary Ann*, for example, combined the names of

Leo Stallman's wife, Marilyn, and Hank Steffens' wife, Ann. In the case of the *Janice Sue*, Louis Steffens—who was single—named the boat after his niece, Janice Sue Stallman, the youngest child of his sister Marilyn (Steffens) Stallman. Glenn Garthe recalled that Louis named the boat to lift the spirits of his sister, who had felt the stress of her fisherman husband Leo Stallman's near misses in storms and ice on Lake Michigan. Three-year-old Janice Sue christened the boat in the fall of 1958. Glenn Garthe remembered the christening: "Uncle Louis took everybody out on the *Janice Sue*, and we had such a load of people on board that

Leo Stallman Sr. in the steel-hulled *Mary Ann*, 1959. The hull of the *Mary Ann* currently is owned by Fishtown resident, Nicholas Lederle. COURTESY OF GLENN GARTHE

the propeller was just digging the bottom up as we backed out the river."[16]

In 2008, at the fiftieth birthday celebration for the *Janice Sue*, Janice Sue Stallman Kiessel rechristened the boat, now owned by Fishtown Preservation. The vessel has passed from the Steffens to the Carlsons to FPS. The *Traverse City Record Eagle* recently referred to the *Janice Sue* as "Fishtown's iconic fishing tug." Indeed she is. With the exception of the years 2003-2009, she has continued Leland's proud commercial fishing legacy.[17]

Fishtown's other steel tug represents a comparatively new phase in Leland's commercial fishing history. The *Joy*—built in 1982 by local fisherman Ross Lang working with George Stevens of Lake Leelanau—was Fishtown's first and only trap net boat. In 1968 Lang and his father Fred had arrived in Leland via Alpena and the Upper Peninsula. Like everyone else, they fished with gill nets for the dwindling numbers of chubs and whitefish until State of Michigan regulations forced their hand. In the mid-1970s the Department of Natural Resources first banned the large-mesh gill nets used to catch whitefish and then closed chub fishing to protect the collapsing population. Brian Price was a college student

The *Janice Sue*'s christening, 1958. COURTESY OF JANICE SUE KIESSEL

Ross Lang with the newly completed *Joy*, c. 1982. COURTESY OF JOY LANG ANDERSON

fishing with Langs at the time. "The state had a moratorium lake-wide on catching chubs," he recalled, "and offered to let those people catching chubs convert to other gear [meaning trap nets]." Then came the state-mandated lottery to distribute a limited number of special chub research licenses. A local newspaper described the event as "more a funeral than a lottery—a breed of men and a way of life on the brink of extinction." Bill Carlson won the lottery, but other fishermen weren't so fortunate.[18]

For those who lost the lottery, trap nets were the most viable option. State regulations had periodically banned and reinstated trap nets, but in the mid-1970s trap nets were viewed as a more environmentally sound option since the "trapped" fish could be culled and protected species thrown back. Trap net boats were completely different from gill net tugs, with a long open section between the pilot house and the stern to facilitate retrieval of the "pot" or entrapment section of the net. The conversion of one boat type to the other was costly and time-consuming. The alternative was to buy a new boat.[19]

Many beleaguered commercial fishermen went out of business, but not Ross Lang. Lang had such a love for fishing recalled his wife, Joy, that he did what needed to be done to keep fishing, including the switch to trap nets. "I don't know of anyone that loved their job as much as he did," she recalled. "It was kind of refreshing to see that sort of thing." During the late 1970s Lang first purchased the aging trap-netter *Seagull* in the eastern Upper Peninsula but soon found he wanted a larger, newer, and more comfortable boat. The high cost of purchasing a new vessel prompted Ross to put his engineering talents to work. Joy Lang remembered "him sitting there with heavy cardboard designing the hull of the boat, getting just the right pitch. And that's all he had. The rest was all in his head." Construction took place in an East Leland airplane hanger with George Stevens welding the steel hull and Lang doing the rest. "The men spent about thirty-five hours per week over four months leading up to the launch in March 1982." Named for Lang's wife in the traditional manner, the *Joy* became the first boat since the *Good Will* to be completely built by local men.[20]

Ross Lang fished with the *Joy* for sixteen years. The boat was his joy and his

The *Joy* and skiff at Fishtown, late 1980s. Ross Lang drowned when the skiff flipped in an accident in 1998.
COURTESY OF JOY LANG ANDERSON

livelihood. He died in an accident on 23 April 1998 while pulling anchors in Platte Bay. Ed Peplinski was working crew that day and witnessed the accident from the *Joy*. "The water was very cold, like forty degrees, but the lake was like a mirror. Nice sunny day, perfect day to do what we needed to do—move the net." Ed was on the *Joy*'s deck to retrieve the anchor lines "that stretch the net out with these eighty pound anchors." Andy Foglesong and Lang, in the skiff, had the job of pulling the anchors off the bottom.

The men weren't wearing life jackets, and when they lifted the second anchor something went amiss. As Peplinski watched helplessly from the *Joy*, the skiff turned on its side, filled with water, and sank. "What I think had happened is that a piece of plywood in the floor of the boat hit Ross in the forehead. Because he just disappeared, and he was a better swimmer than Andy or I put together." Lang's close friend and one-time fishing partner, Alan Priest, was working on the *Janice Sue* when the news came over the radio. "They didn't know who it was at first," he recalled. "Then they said it was Ross, and I just kind of lost it. Because you never expect it to be Ross. Everybody called him bulletproof. It was a shock."[21]

Fishtown's two fishing boats, the trap-netter *Joy* and the gill-netter *Janice Sue*, 2006. PHOTO BY KEITH BURNHAM

The *Joy* on her first run of the year, 2011. PHOTO BY AMANDA HOLMES

With Lang's death, Fishtown lost an excellent, innovative, and energetic fisherman. The *Joy* continued to be based in Leland, however, initially with the Manitou Fish Company and currently with Fishtown Preservation. Like the *Janice Sue*, the *Joy* was not used from 2003 to 2009, but thanks to restoration efforts undertaken by FPS, both boats are fishing once more. As the *Joy* motors out of the harbor, she is a testament to the man who built and captained her, Ross Lang. He would be glad she is still a working boat.

Early mail boat captain John Swenson made his runs between Leland and North Manitou Island in a Mackinaw boat that probably doubled as his fish boat, 1903. PHOTO BY FRED A. SAMUELSON, COURTESY OF LEELANAU HISTORICAL SOCIETY

Chapter 7

"Special Link to the Mainland": Fishtown's Manitou Ferry and Mail Boat

> "A south wind took the ice away from the Leland Harbor Wednesday, and Captain Grosvenor got out with his mail boat and made a run to the North Island. During the night the wind shifted and now there is ice for several miles out. It will probably be some time before he can make the return trip."[1]
>
> – *Leelanau Enterprise*, 1930

When third-generation ferry captain Mike Grosvenor sits at the helm of Manitou Island Transit's *Mishe-Mokwa*, he represents a family tradition. The Grosvenor family has operated the ferry and mail boat for four generations, working from a base on the shore of the Leland River at Fishtown for nearly a century. Today's 70-foot passenger ferry would dwarf the diminutive "winter boats" Mike's grandfather, Tracy Grosvenor, once took across the treacherous ice-filled waters of the Manitou Passage. Now Manitou Island Transit doesn't even run in the winter. Most trips take visitors for a tour of South Manitou Island. But there was a time when the ferry served a resident population on the islands. The Grosvenor's mail boat once carried vital supplies and mail to the tiny community on North Manitou Island throughout the year. It was not hyperbole when the *Detroit News* carried the headline, "Defies Death to Get Mail Through: Capt. Grosvenor Battles His Way To Ice-Bound Isle Twice a Week." George Grosvenor, the next-generation captain, learned well from Tracy's example. He was "courageous to a fault," Mike recalled of his father. "He had this idea that he was [the islanders'] special link to the mainland. He had this obligation that he felt to the extreme that he had to make these trips."[2]

The Manitou Islands, especially the geographically closer North Manitou Island, have played an important role in the history of Leland and Fishtown. The islands were populated before the mainland, and many early European American settlers came to Leland via the Manitous. Among them were the fishing families Clauss, Firestone, Maleski, Carlson, Johnson, and Buckler, and the Grosvenors with the mail boat and ferry. Leland was the closest mainland port to North Manitou Island and the staging ground for a vital supply link to the island community, which peaked in population at about two hundred between 1910 and 1920. Generations of mail boats doubling as ferries traveled the Big Lake between the two as the island's enterprises evolved. Significant ventures included cordwood production, logging, commercial fishing, farming on cut-over lands, resort colony, corporate farming of the Manitou Island Association, and private hunting preserve, before the island was subsumed by the National Park Service as part of Sleeping Bear Dunes National Lakeshore.[3]

During the nineteenth century, men who had their own boats—including

Early postcard with (on left) Mackinaw fish boat, the type of boat also used as early mail and supply boats; cattle to be shipped to or from North Manitou Island on the beach below Fishtown; and an early passenger launch similar in design to John Paetschow's *Lawrence*. COURTESY OF MIKE BROWN

fishermen—provided an impromptu ferry service across the Manitou Passage, by some accounts making a tidy profit. The earliest reference to a mail carrier is Andrew Paetschow, who had emigrated to North Manitou Island from the Mecklenburg region of Germany. When the weather was favorable, he sailed twice weekly to Leland. Perhaps the earliest boat built expressly for the mail contract was the *Manitou*, completed in the fall of 1897 for Nels Carlson, who in 1904 would move his family off the island to Leland and become the first Carlson to fish out of Fishtown. Nels had received the mail contract earlier that year and hired boat builder Christian Telgard to construct a special-purpose vessel that the *Leelanau Enterprise* described as "an elegant new mail boat."[4] Other fishermen briefly held the mail contract, among them Alvin Price and John Swenson.

By 1903 Swenson was assisted by John Paetschow, son of Andrew, who would take over the mail contract on his own in 1904 and hold the position until about 1920. John Paetschow's new vessel, the *Lawrence,* was the first boat crafted as both a passenger and mail boat. She was longer in length than fish boats built at the same period and similar in design to small gas-powered launches that ferried passengers around protected inland lakes and harbors. These were smaller versions of the larger passenger steamers that plied Lake Leelanau, such as the *Tiger* and the *Leelanau.*

The *Lawrence* (right) in the river at Fishtown, c. 1912. COURTESY OF LEELANAU HISTORICAL SOCIETY

The new ferry building, freshly painted in white, with freight piled on the curving dock that followed the river shoreline east to the dam, 1930s. COURTESY OF LEELANAU HISTORICAL SOCIETY

In 1909 Paetschow hired boat builder Christian Telgard to add a covered cabin to the *Lawrence*, which provided protection for passengers, mail, and freight. Paetschow began operating his North Manitou Ferry Line during the peak era of Cottage Row, North Manitou's resort colony, and occasionally brought excursions over to Leland from the island.

Tracy Grosvenor, John Paetschow's brother-in-law, began assisting on the mail runs in 1917. A member of the Grosvenor family has run the mail boat and ferry service since, each with a period of mentoring before assuming the helm on his own: Tracy Grosvenor, 1921-1952; son George (known as Sonny), 1953-1983; and George's son, Michael, beginning in 1984, after the two had partnered together for some twenty years. Although Mike Grosvenor still pilots the boat, he has conveyed Manitou Transit Inc. to his children. Each generation inherited the innate ability to run a boat, what Mike describes as "something that comes up through the bottom of your feet from the boat. It talks to you—this current's got your bow the wrong way, or you've got windage here, and you need more power. It's just something that seems to be an instinct."[5]

The family hasn't always worked on the water. Tracy Grosvenor's parents came to North Manitou Island in 1910 where his father worked as a sawyer in the burgeoning west-side timber industry at the island town of Crescent. When Tracy joined the North Manitou Ferry Line, his employer was the Manitou Island Association (MIA), a quasi-corporate entity of Chicago-area businessmen and professionals, successor to the earlier Manitou Island Syndicate, which by 1923 owned most of North Manitou Island and controlled the island's social and economic life. The MIA focused its operations on agriculture (especially free-range cattle and orchards), timber harvest for the maintenance and construction of island buildings, and sport-oriented recreation: a small deer herd was introduced in 1926. All of these enterprises required a dedicated building and dock space in Fishtown to accommodate passengers and freight. Within a decade after taking over the business, Grosvenor oversaw the construction of a new warehouse and post office in Fishtown for use by the North Manitou Ferry Line. The MIA probably funded construction.[6] Completed in 1928, the building anchored the east end of Fishtown and gave the mail boat a tangible footprint in the evolving landscape of boats, docks, and buildings.

Tracy's son, George, recalled that "a small compass was his father's only piece of navigation equipment. 'He sailed by instinct more than anything else; he knew the waves.'" Writer Karl W. Detzer, a resident of Leland, captured the essence of

The former North Manitou Ferry and Mail Boat Warehouse, currently the Nicholas and Susann Lederle House, 2010, prior to significant remodeling to the plans of architect Sarah Bourgeois of Traverse City in 2012. Portions of the original building remain in the lower floor. PHOTO BY LAURIE KAY SOMMERS

Grosvenor in his special 1931 report to the *Detroit News*. "Capt. Grosvenor is the island mail carrier, and there is hardly a man on the whole rocky length of the fishing coast who does not know and respect him. There, where never a hull, or a spar, or a streak of smoke marks the steamship lanes after January 1, he sails his lonely course, with no thought of calendars. The greatest hazards lie along the mainland shore. Westerly winds pile the ice up there, all along the beach, and Leland Harbor mouth is a treacherous port at best. Ice-sheathed pilings thrust out at both sides of the narrow entrance, and a swift current rushes outward to meet the lake. There, buffeted by waves, yanked by current with the sharp jaws of the ice packs snapping, he must steer skillfully and truly to bring his boat to dock." [7]

Like the fishermen of Fishtown, the mail boat captains relied on their skill, the grace of God, and above all on their boats. John A. Johnson, the Omena-based builder of so many of the Leland fish tugs, built Tracy's first new mail boat in 1919

This photo accompanied Karl Detzer's 1931 *Detroit News* article with the caption, "The Needle's Eye—with a handbreadth to spare between the boat and the icy piers; Capt. Grosvenor brings his boat into Leland's harbor."
COURTESY OF LEELANAU HISTORICAL SOCIETY

(most likely the *Bob*, Tracy's first boat after the *Lawrence*). A succession of vessels followed: *Manitou*, *Shirley*, *Fern-L*, *Manitou II*, *John A.* and the converted fish tug, *Smiling Thru*. The *Shirley* and the *Fern-L* were named for Tracy's daughters; the latter was a converted Coast Guard boat from North Manitou Island. Johnson crafted the *Manitou II* and the *John A.* in Fishtown and the latter was named for him. He originally built the *Smiling Thru* for Northport fisherman L.J. Strayer; she was Tracy Grosvenor's last boat.

Grosvenor initiated the practice of using a summer and a winter boat. In a seeming paradox, the smaller boat was for winter. Since there is no natural harbor at North Manitou Island, a smaller-sized boat was easier to pull on shore or launch during the stormy winter months. The diminutive *Manitou* was the tiniest of the winter boats, its scale seemingly incongruous in the face of winter

Tracy Grosvenor's 26-foot winter boat, the *Manitou*, featured an open "fantail" stern design, 1930s.
PHOTO BY ERHARDT PETERS, COURTESY OF LEELANAU HISTORICAL SOCIETY

Tracy Grosvenor's summer boat, the *Fern-L*, leaving Fishtown, 1930s. PHOTO COURTESY OF LEELANAU HISTORICAL SOCIETY

gales and perilous ice floes. Karl Detzer captured the dichotomy of Tracy's winter and summer boats. "North Manitou is a busy place in the summer months," he penned. "A great orchard stretches over hundreds of its fertile acres. Traffic is heavy on the daily trips of the island mail. The little mail boat, *Fern-L*, plying those pleasant waters, rarely meets obstacles."[8]

"But when the ice closes in on the lakes," continued Detzer, "and the service is reduced to two days a week, the element of uncertainty and the tang of danger creep into this prosaic business of transporting the mail. Grosvenor puts up his summer boat, which is nearly 40 feet long, and gets out his little ice breaker, a nameless craft 26 feet over all, staunchly constructed to meet the buffets of the floes in motion."

In December 1928 the *Leelanau Enterprise* reported that the mail boat had been reinforced with metal to protect against ice, a technique also used by fishermen prior to the adoption of steel-hulled vessels in the late 1950s. Despite these improvements, the winter crossing was still a perilous business. "Twice a week, sometimes with a packet of mail he easily could carry in his hat," Detzer wrote, "Grosvenor launches his boat on the icy beach of Manitou. Between him and the mainland are ice packs, patches of open water, treacherous hidden bergs. With an uncanny skill, he always has managed to get through. Given up for lost a dozen times, he goes about this job of his with less fuss than the average carpenter getting out his tools."

During the winter months, the mail boat was the lifeline for the small, otherwise isolated North Manitou community, bringing groceries from the Leland Mercantile, medical supplies, a doctor in times of emergency, and the comfort of the mail. "Repeatedly," Detzer wrote, "when the most skillful sailors on the lake insist that a trip would be suicide, with a [Coast Guard] surfman or a fisherman or two to help him, Grosvenor will set off with a country doctor clinging to the handgrip on his deck."[9]

Tracy Grosvenor didn't miss many trips. His son, George, continued that philosophy. After George and Mike became partners, the two would disagree about the risks of the winter boat runs, Mike telling his father that "the magazines and newspapers and bottles of milk that we're taking over there through these gales are just not worth it. I always told him that we should limit the mail contract to eliminate winter months, but he felt this obligation. If he couldn't run the boat over to the islands, he would charter an airplane in Traverse City to drop the mail out of the window over the farms over there, or he would walk

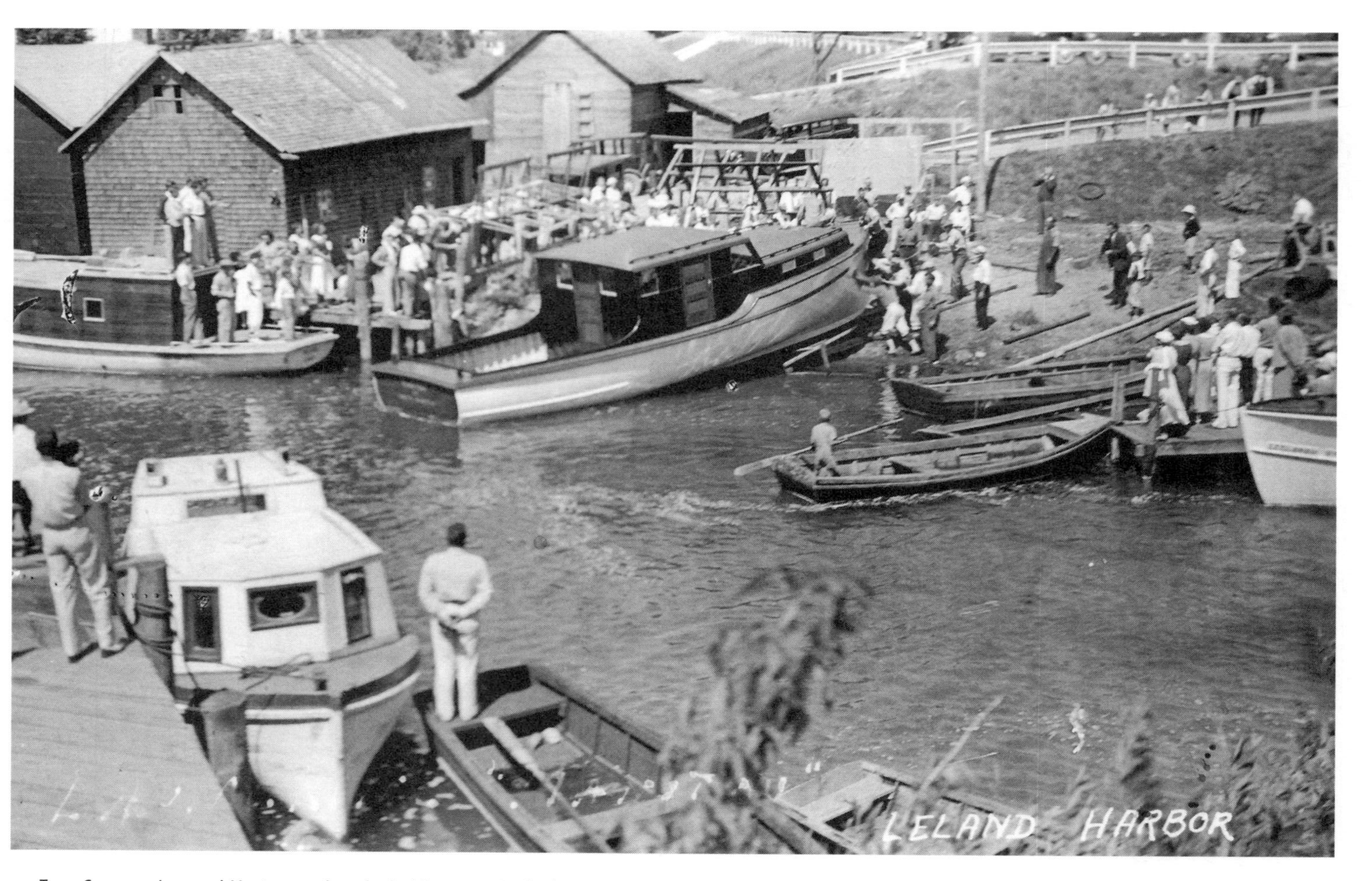

Tracy Grosvenor's second *Manitou* was launched with community fanfare in 1936. COURTESY OF LEELANAU HISTORICAL SOCIETY

from Sleeping Bear Point across the ice to the islands to deliver the mail." Mike finds it ironic that his own kids "look at me as being overly dedicated to, 'Oh, we can get this trip in. We can't afford to lose this trip.' They look at me and say, 'Maybe you did in your era, but we don't go out in that kind of sea.' The bar was set very high back in the older days. In each generation it gets lowered a little bit, to where now the kids are probably being pretty sensible."[10]

The act of piloting a boat in the gales and ice floes of Lake Michigan created a special bond between fishermen and the mail boat captains. As Mike Grosvenor described it, "Through the generations, that whole Fishtown community seems to be a brotherhood. To be part of that brotherhood is to have those experiences that you share. It was a tight community; you were all concerned over one another. You made sure everybody was home when they were supposed to be home. There

"Helping Tracy get in" was Detzer's caption, 1931. COURTESY OF LEELANAU HISTORICAL SOCIETY

was no communication then, you were just watching out for each other. If there was a mechanical breakdown, they were all down there, getting the other guy's fish tug going. That passed on to my father's generation."[11]

Karl Detzer wrote about this bond in 1931. "In the winter months, when the ice packs are grinding and the north wind roars, islanders and mainlander alike go down to the lake with Tracy Grosvenor, and fight with him to get his boat safely from port to port. Fishermen, coastguards, light keepers, sailors, there are few able bodied men on the coast who at some time or another haven't dragged a rope, or held aloft a flare to guide the island mail."[12]

Just after World War II, on George Grosvenor's first trip out with Tracy, the boat "became locked in a grinding mass of ice that threatened to destroy the craft and take father, son, and crewman Harry Garthe to the bottom." The *Leelanau Enterprise* described how "they were trapped thirteen hours, but at dawn the ice broke enough to allow two fishing boats to force their way through and rescue the mail boat."[13] And the bond was reciprocal: the mail boat went to the aid of stranded fishermen. The *Smiling Thru,* for example, was part of the search party for Will and Pete Carlson after the *Diamond* burned in 1941, and it towed the new hulls of the *Janice Sue* and *Mary Ann* from Marinette, Wisconsin, to Leland in 1958.

The *Smiling Thru* was Tracy Grosvenor's last boat and George Grosvenor's first when he took over the business on his own in 1953. The *Smiling Thru* still carried primarily freight: during the 1950s the dock near the Fishtown warehouse would be piled with lugs of cherries coming off the island and tons of deer feed headed to the Angell Foundation hunting preserve on the island. When George Grosvenor started sailing with his father in 1946, tourist traffic was only seven hundred passengers per year. Today, it is about ten thousand per year. The succession of boats illustrate this transition, from the *Smiling Thru*, a fish tug converted to a passenger boat (1950-1965), to the newest and largest boat in the ferry's history, the *Mishe-Mokwa*, purchased in 1981.[14]

The modern harbor of refuge has replaced the old battered seawall, and the ferry no longer braves the winter ice and gales. Gone are the days when Tracy Grosvenor would stagger into the port of Leland, "an hour overdue, his decks coated with ice." On that stormy winter day in 1931, Karl Detzer witnessed how "fishermen and townsmen had been scanning the horizon, wondering what had detained him. They could see the waves rolling in great black ridges along the skyline, and they knew that the little boat was fighting the seas. Now, safe ashore,

The *Mishe-Mokwa*, moored adjacent to Manitou Island Transit's latest ferry building, 2010. About 1960 the Grosvenors moved their operations from their historic warehouse site near the dam to the present location at the river's mouth. PHOTO BY LAURIE KAY SOMMERS

he climbed out, the thin mail sack under his arm. Someone said, 'Hard trip?' 'No,' Grosvenor shook his head, and the ice cracked on the collar of his jacket. 'Not, not a hard trip. Just a little sloppy, that's all.'"[15]

Yet even without the drama of winter ferry runs, it can still be a tricky business bringing the 62-foot *Mishe-Mokwa* into port during rough weather. "You have to do everything right," Mike Grosvenor explained. "You probably have to take one of these breaking seas as you're turning into the harbor, which, with that boat, can roll you over pretty dramatically. You have to prepare your passengers and balance the boat, and then the rest is timing and luck. Of course you have to add your own skill into that. You can only pace yourself to find a nice spot between the seas coming in for so long. Once you're there, you have to take what the lake gives you. Sometimes you're lucky, and sometimes not so lucky." A true test of the captain's skill has always been negotiating the harbor entrance. In George Grosvenor's day, fishermen and ferry captain would gather after docking their boats in a particularly bad storm. "Everybody had to make that harbor entrance," Mike recalled. "There would always be this kibitzing about who had the roughest ride, about whose boat handles better."[16]

Mike Grosvenor, like Tracy and George before him, feels the special bond with those who pit their wits against Lake Michigan. While the earlier generations interacted more with commercial fishermen, today the greatest number of "captains" in Fishtown run charter boats. "We all sort of come in about the same time together," Mike explained. "Of course the *Mishe-Mokwa* comes in with a bar. So when she ties to the dock, everyone who has been out there on that day will migrate down to the *Mishe-Mokwa*. We'll sit around and bitch about the weather together, have some drinks, tell lies. There's a camaraderie. There's a feeling that you're obligated to watch out for each other and take care of each other. It's a nice security blanket to have that team behind you. It becomes a long-lasting and tight social network."[17]

Chapter 8

Brotherhood of the Lake

> "Going out in the morning, seeing the sun rise on the lake, it's an incredible feeling. Nobody tells you what to do. We're all brothers out there, having a good time, you know? It's a unique job."[1]
>
> – Jim VerSnyder, 2008

JIM VERSNYDER HAS A GIFT FOR CONVEYING THE ESSENCE OF THE FISHING brotherhood to landlubbers who have never set foot in a fish tug. He married into the Steffens family and cut his fishing teeth in the 1960s with Pete and Bill Carlson aboard the *Janice Sue*. He loves the water. Loving the lake makes it worthwhile to work in such a dangerous occupation. Unlike old-timers Pete Carlson and Roy Buckler, VerSnyder wasn't born to it. But they shared their love of lake and fishing. Once, when an aging Pete was asked if he would still choose commercial fishing as an occupation, he replied, "Well, being born and raised on Lake Michigan, my grandfather and father being a fisherman, I wouldn't want any other kind of life. If I had to do it over, I'm sure that I would choose fishing again."[2]

Lake work offered the pure beauty of dawn on the water and the camaraderie born of weathering a storm. In the crucible of accumulated experience, fishermen learned the crucial skills of their job: setting and lifting nets; the best way to repair a torn net or dress a fish on the way back to shore; predicting the weather; deciding whether to ride out a storm or seek shelter; navigating in a rough sea; understanding the ways of fish and the best places to set a net; the names of

Jim VerSnyder reminiscing about his fishing days from the *Janice Sue*'s cabin during Fishtown Preservation's Shanty Aid, 2010. VerSnyder now does shore work for Carlson's of Fishtown. PHOTO BY LAURIE KAY SOMMERS

fishing grounds. "Once you learn how to do it," Jim VerSnyder observed, "You're your own boss. You're working with a guy that's paying you, but you go out there, you just pick your fish and you laugh and joke. It's hard work, but it's out in the elements." It gets in your blood.[3]

Not everyone has the temperament or the constitution to make it as a commercial fisherman. Clay Carlson grew up during the 1970s and 1980s, the son of Bill Carlson and the grandson of Pete. As a boy, he crewed on the *Janice Sue*, absorbing the stories of his elders and learning the value of hard, rugged work. His mentors were fishermen who had spent their lives on Lake Michigan, men with strong, thick hands and skin creased from sun, wind, and cold. Men like his grandfather, who survived the wreck of the *Diamond* in 1941. "The work was harder back then, the men were harder back then," Clay observed. "There was no such thing as a computer, all these modern conveniences we take for granted. They were a much tougher breed, I'll tell you. Much tougher."[4]

Jim VerSnyder also marveled at the stamina of the old-timers. "They were

George Cook setting gill nets from the open fantail stern of the *Why Not*, c. 1920. PHOTO BY ERHARDT PETERS, COURTESY OF LEELANAU HISTORICAL SOCIETY

right out in the elements, just spinning the nets in the cold weather. There was ice on their hands—I mean, they were tough. To this day I'm still in awe of what those guys did, without the equipment, on open back decks. Going out there on those old wooden boats with gasoline engines was dangerous as heck. They were really incredible guys."[5]

Today's gill net fishermen motor out of the river at Fishtown in an enclosed cabin, with the captain manning the helm. By contrast, the old wooden boats, as remembered by Glenn Garthe (grandson to Henry Steffens), "were steered by a tiller, so the guy 'wheeling' the boat would have to stand out in the weather and steer it with his feet. Across the back of the boat you see a handrail—that's what they hung onto. That's where the ice formed in the winter. They're hanging on, they're going through the seas and rocking and rolling. These guys were pretty tough. But eventually they put cabins on the back, because they started to get older, like we do, and then, 'I can't take the cold anymore.'"[6]

Bill Carlson recalled one of Pete's favorite axioms. "[He] used to always say, 'In the old days we had wooden boats and iron men, and now you have iron boats.'" Pete reportedly also had an iron stomach and was never seasick. Jim VerSnyder was less fortunate. He described working inside the *Janice Sue*'s cabin as like being in a "dark milk jug." The combination of diesel fumes, fish guts, and the boat rolling in a rough sea "was a little tough to get used to. I got sicker than a dog."[7]

Pete Carlson could be a merciless tease with a queasy crew member. Jim VerSnyder recalled how Pete would say, "'Hey VerSnyder, you want a piece of pork on a string?' I had the dry heaves, and he said, 'Take a swallow of that piece of pork string, pull it out, and it will clean you right out.' Then he would light up an L & M cigarette. When I was fishing with Terry Buckler they would have their cigars going. That is one of the worst things you can sniff when you are out on the lake and you are feeling a little queasy. They were always poking fun at you about getting sick, and they thought it was funny."[8]

Clay heard another story about Pete's remedy for a seasick crew. "It was sort of a wavy day. By the end of the day you had about three inches of slop, which was mainly fish guts and water mixed together, sloshing back and forth on the bottom at your feet. Some guys probably weren't feeling very good. So my grandfather decided to take his coffee cup, and he bent over and scooped up a cupful of the

Fishermen working inside their fish tug, date unknown. PHOTO BY ERHARDT PETERS, COURTESY OF LEELANAU HISTORICAL SOCIETY

slop. He had some sort of remark, like 'Bottoms up,' and he guzzled it right down. That was sort of a practical joke to make sure he got everybody past the point of feeling a bit sick, making sure they got sick."[9]

Lake work was not all wind, waves, and slop rolling at your feet. VerSnyder also remembered days of flat water, fish-filled nets, and sheer fun. "We went over to Grand Traverse Bay and fished for whitefish and chubs. Nobody's done that for many years now, because that was a DNR research permit." VerSnyder remembers the bay as "just a wonderful place to fish. You're out of the wind most of the time. The size classes of fish were just amazing. Whitefish—we get jumbos down to the little ones. The chubs—there were maybe eight different species of chubs over in the bay. It was just unbelievable fishing, and we had the time of our life. Then we had to get out of there and come back to the real world, where there is no lee. You go out to the west of the Manitous and get beat up all the time."[10]

From the Manitou Islands to Grand Traverse Bay, lake work takes place within a vast underwater geography of fishing banks, each with their own name and characteristics. Folklorist Michael Chiarappa noted the importance of these "customary locations or home waters," which for multiple generations have linked Fishtown's traditional fishing territories to fish tugs and shoreside facilities.[11] These are the banks that local fishermen knew—and still know—like the back of their hand. In years past, boats freely roamed Lake Michigan following the fish. Since the 1960s, however, the Michigan Department of Natural Resources has assigned fishing licenses to particular zones of the lake to prevent overfishing, and the catch is limited to a quota per license. Some historic fishing grounds are now off limits, either by quota, Indian fishing rights rulings, or the establishment of special trout sanctuaries, such a those between the Fox Islands.

Generations of fishermen learned the locations of the fishing grounds and became masters in the habits of fish. "They had a sixth sense," said Brian Price, "an ability to read the water and the currents" without the benefit of sounders, charts, and electronic equipment. Jim VerSnyder recalled Pete Carlson's sixth sense. "'Let's go down here about a half of a mile,' he would say. 'There's a hook in the bank down there, and the chubs always seem to hang on that hook.' Sure as heck, the next time we go out, we'd have a load of chubs. Now it's easy, with the electronic equipment. I can't imagine doing what they did without it. And the thing about Pete was, he never even kept a log. He never wrote in a book or anything. It was all in his memory. He just lived and breathed it."[12]

Fishermen in the age before radar were expert navigators. Their skill with a

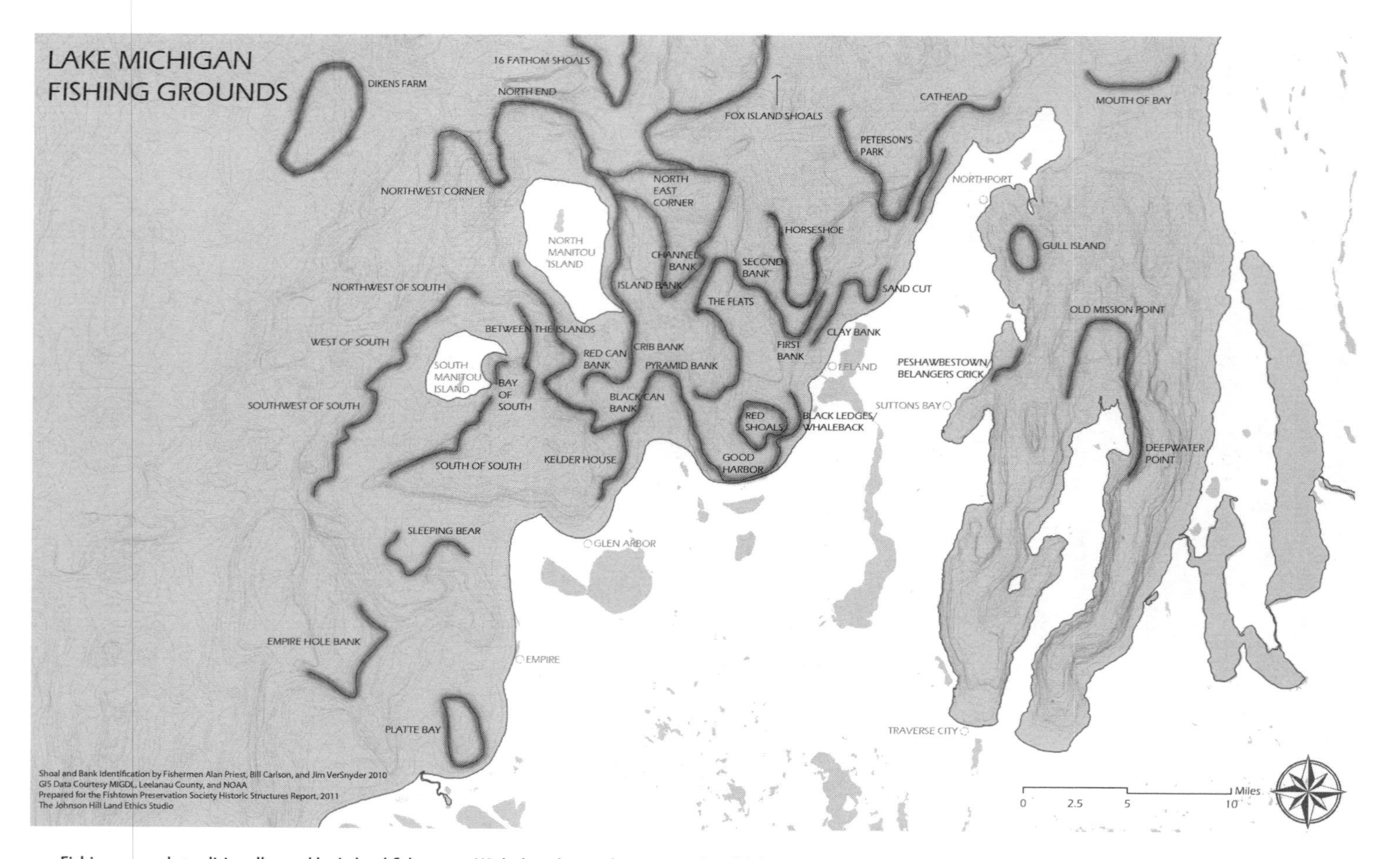

Fishing grounds traditionally used by Leland fishermen. With the advent of state-mandated fishing zones, some of these fishing grounds are now off limits. MAP BY JOHNSON HILL LAND ETHICS, ANN ARBOR

compass was commonplace for the time but legendary today. VerSnyder still marvels at their ability to find a net buoy even in a dense fog or storm: "These guys were really good at ranges. But if you had fog, you're pretty much going by your compass. So you're running so many minutes—an hour and forty minutes or two hour—and then you stop and you look for the buoy. Well, if it's foggy out there, trying to pinpoint that little stake that far out, you've got to be really wheeling a true course. When we got into the radar, they had reflectors they'd put on the buoy staffs, so when you got close, it would pick that up on the radar. They'd run right up to it, whether it's foggy or whatever."[13]

Buoys mark the location of both gill and trap nets, the latter used in Fishtown

Buoy viewed through the window of the *Janice Sue*, 2010.
PHOTO BY MEGGEN WATT PETERSEN

only since the 1980s. Historically, fishermen also used pound nets. Each net required a different boat design and skill set. Trap-netters, for example, have open decks best suited to fishing in calmer waters. Gill nets are easily moved, but trap nets remain in one location, held down by anchors on the lake floor. Brian Price, who crewed on the trap net boat, *Joy*, described setting trap nets as a shell game. "You're always trying to outsmart the fish, even more than with the gill net fishery. When you hit it right, trap nets can be wildly successful. You're starting to pull the net up, and as the fish rise through the water—maybe the net's in 80 or 90 feet—you see the water start to boil because they're expelling air from their air bladder. It's just fun. Sometimes 6,000 pounds is not unheard of in a trap net, and it's like hitting the lottery."[14]

Earlier generations of fishermen used pound nets, another impoundment technique that disappeared with the arrival of the sea lamprey. "When I went to fishing," Roy Buckler recalled, "they used those [pound nets] for years and years down here. They were hung on sticks and had a lead, a pot, and a trap. [Fish] would follow the lead out from the beach and get caught in the trap." Pound nets required calmer waters and comparatively shallow depths in which to pound

Joy captain Jerry Vanlandschoot of Munising and Adam Kilway dip whitefish from a trap net, 2010. PHOTO BY MEGGEN WATT PETERSEN

the stakes that secured the net, hence the name pound net.[15]

In the summertime, Buckler would lift his pound nets and get "whitefish by the tons" as they moved offshore to feed in shallower water. "Yeah, just empty the pot and bail them out," he recalled. "When the fishing was good you'd go every day, and you had about five nets. You could get a hundred fifty pounds out of a gill net in a day or you could get a thousand pounds out of a pound net. It didn't last but two and a half months, and then it was over again. It was easy fishing—no nets to spread."[16]

Gill net fishing coincided with pound nets but had a much longer season with different techniques and challenges. Dick Carlson (born 1934) remembers seeing his uncles setting nets from the open back of the boat. The men would stand on either side of the carefully packed net box, corks on one side and leads on the other, and spread the nets apart until they dropped. "They'd just fling the corks and the leads out," Dick recalled. "The weights would drop in the water and the corks would stay on top."[17]

When Jim VerSnyder started in the 1960s, they set gill nets with their bare hands. "You could actually feel the meat on your hands kind of getting warm and smelling a little bit, because the nets were burning your hands. Alewife belly bones would go through there and cut you up. Finally, somebody says, 'Why don't we wear some rubber gloves?' So we started wearing rubber gloves, and that really made the hand issue a lot better deal."[18]

Fishermen lifting a pound net with a skiff in the shallower water off shore, 1930s. PHOTO COURTESY OF LEELANAU HISTORICAL SOCIETY

Roy Buckler (right) and unknown fishermen setting a gill net, c. 1930. PHOTO BY ERHARDT PETERS, COURTESY OF LEELANAU HISTORICAL SOCIETY

Alan Priest and Albert Gunderson setting gill nets in the *Janice Sue*, 2010. PHOTO BY MEGGEN WATT PETERSEN

Since the 1890s, fishermen have used mechanical lifters, but a net full of fish still involves teamwork and what Jim VerSnyder calls "hand work." The captain has the tricky job of positioning the boat as the nets come through the lifter. Other crew members may "pick" fish from the net, stow the nets in a box, or dress fish. When Clay Carlson crewed on the *Janice Sue*, his job was to pick fish and then rope the net back in the net box so it could be reset without tangles. "Which meant I was the closest one to the window on the one side of the boat," he explained. In the winter, it was so cold his hands "were like claws" by the end of the day. The boat had a window on one side "where the winter wind would be whipping in. Then you had this wood stove on the other side of you, blasting heat. So it was like one side was on the cusp of frostbite, and the other side made you wish you had a sleeveless coat on."[19]

The tug's stove functioned for more than needed warmth on cold-weather runs. "We used to cook almost every other day when we would go out," Jim VerSnyder recalled. "We liked trout on the boat. Pete is the one that turned me onto that." The men topped the filet with margarine, salt, and pepper, and then wrapped it in paper soaked in a bucket of water. "When you put it on top of the stove it's like

Chub caught in a gill net is pulled up with *Janice Sue*'s mechanical lifter, 2010. PHOTO BY MEGGEN WATT PETERSEN

Albert Gunderson using the dressing block, special knives, and lifting table aboard the *Janice Sue* to dress chubs on the trip back home, 2010. PHOTO BY MEGGEN WATT PETERSEN

you are steaming your fish. Man, that was incredible tasting. It was some of the best fish I have ever eaten. In the summer, some guys just cooked them on the boat's manifolds. They'd wrap it up in aluminum foil and put it on a manifold. We would do the same thing."[20]

In years past, the men fished until ice made the harbor impassible, so a stove was an indispensible part of a tug's equipment. The boat carried extra wood or coal and emergency canned goods. If fishermen were caught in a storm or ice floe, they could survive by cooking fish and hovering around the stove to keep warm. "On the cold days we used to have the floorboards lift up," VerSnyder remembered, "and we would have a big tin and some extra coal down there because the coals would burn no matter what. Coal would last a lot longer than wood."[21]

Wintertime fishing presented more challenges than merely staying warm. Since their livelihood was based on the size of their catch, fishermen worked deep into winter, braving treacherous ice floes until ice clogged the harbor entrance. In the early spring they were anxious to get back to work; as soon as the ice began to break, they would try to head out. Many times they would "buck" ice between the harbor and open water. For Jim VerSnyder, bucking ice in the steel-hulled *Janice Sue* was "the god-awfulest noise you ever heard. It is crunching and vibrating and the diesels revving up. The bow of the boat is going right up on top of the ice. When you got up on top she only goes so far, and you can not get her out of the water. She will sit there. Pete would turn the wheel right and left and blow a lot of water up to get the water washing all the ice behind her and out of the way, so you had room to back up and buck it again. Sometimes it would take a couple days to bust through. Once you got out of the channel the current would start opening it up, and you could get in and out."[22]

The old-timers, in their less durable wooden tugs, used pike poles to dislodge the ice or push the boat. "I cannot imagine bucking ice with a wooden boat," VerSnyder admitted. "It was scary enough with a steel boat. You could see the hull shifting. I did not like bucking ice at all. It was kind of spooky."[23]

Using pike poles as the *Nu Deal* and *Diamond* buck ice, 1930s. COURTESY OF LEELANAU HISTORICAL SOCIETY

Ice could move unexpectedly, trapping fishermen at sea, or keep them landlocked and unable to lift valuable nets and catch. Dick Carlson recalled a near miss with his father, Gordon, and Pete Carlson, in the fish tug *Good Will.* When the two left the harbor, the weather seemed fine. "But the wind came up, and they could see that the ice floe was breaking loose and coming down the lake." The men soon realized that they were trapped in a pincher of moving ice. "They got to the edge of the boat," Dick recalled, "and my dad was waiting for this ice to come in and just crush the bottom of the boat." Providentially, the floe merely popped the boat on top of the ice. One of the men walked over the frozen surface to summon help from the Coast Guard ice breaker, *Mackinaw.* "The ice breaker has a propeller in the front that sucks the water out from under the ice and breaks it," Carlson explained. "My dad said there were ice chunks big as the boat flying all over the place. Then, of course, they followed the *Mackinaw* back into Leland Harbor."[24]

Fishermen took chances with the lake because it was their livelihood. "We always thought we could beat it," VerSnyder recalled. "We would get a weather report on the way out and 'Yeah, I think we can get her done.' Sometimes [the storm] would come through a little quicker than it was supposed to and you are caught in it. That happened a lot—getting caught in bad weather." Wives would worry and wait. VerSnyder's wife, Wendy, granddaughter of a fisherman, would be at the docks with the kids. "They would be shaking their heads, 'What are you doing this for?'"[25]

Lake Michigan is not a place for the fainthearted. A rogue wave, a slippery deck, and a fisherman faced a watery grave. Norman Price almost drowned in 1928 while setting nets with his father, Oscar. "The boat was running against a heavy sea," reported the *Leelanau Enterprise*, "and in some way Norman fell overboard. He came up some distance behind the boat and luckily managed to get hold of the net that was trailing behind. This is all that saved his life, as with his heavy fishermen's clothing and hip boots he would have been unable to keep afloat while help was coming. Oscar Price, his father, got hold of the end of the net that was still on the boat, while Lester Nedow, who was with them, reversed the engine. By this time Norman was hanging to the net about 300 feet behind the boat. The other men reached him and with great difficulty succeeded in getting him on board." The fortunate younger Price escaped with nothing worse than a severe chill.[26]

The Big Lake reserves the height of its fury for the infamous storms of November.

Roy Buckler and the *Irene* arriving safely back in port, 1930s. PHOTO BY ERHARDT PETERS, COURTESY OF LEELANAU HISTORICAL SOCIETY

A November blow sank the *Carl D. Bradley* off Beaver Island in 1958, and the *Edmund Fitzgerald* in Lake Superior in 1975. The historic "Big Blow" or "White Hurricane" of November 1913 swept away Leland's bridge pier, along with similar breakwaters in Milwaukee and Chicago. If fishermen from Fishtown rode out that infamous storm, they left no stories behind. Later generations of fishermen did share their tales. The worst storm of Roy Buckler's long career was a 1940 November blow while on the *Irene* fishing for trout off North Manitou Island. On 10 November, they set their nets before dark, laid over off the island for the night, and lifted the following afternoon. "When the wind started to blow—I would imagine about twenty-five miles when the first blast came through—we started for home."

Twenty-five minutes outside of Leland, they were forced to turn around. "It was so bad, raining and half snow, that we didn't think there was a chance that we'd ever get in the harbor." They returned to North Manitou, twice forced to lift anchor and move as the wind howled to seventy knots. Finally, they headed to the east side of the island, where the snow was so thick it took them nearly four hours to find the dock, let alone get a line on the piling. "Then we were in good shape," Buckler recalled, but "one thing that bothered everybody was the storm took the telephone line and nobody knew where we were." The next day, just two miles off the island, "the sea was so big you went down in the valleys and you couldn't even see the island, so we turned around and went back." The weary crew and fish tug *Irene* finally made it home after four days at sea. "The next trip out, I got about two miles out on the water, and when the wind started to blow a little bit, I didn't have nerve enough to keep on going. After about six months, I felt safe on the water again. Yeah, what a life."[27]

Jim VerSnyder has his own November tale from the 1970s, when it was "just flat-out howling on the northeast channel bank. It was blowing so hard we just untied the nets in the middle of a box, and it was a hell of a ride home. It was about as bad as it gets." The other Fishtown rigs, with the Langs and Bucklers on board, were weathering the same storm on the west side of North Manitou Island. "The Langs took water through the side of their lifter door," VerSnyder recalled. "There was solid green coming in. Old Fred Lang took a knife and cut his own nets. There was no untying or anything. He was that scared. They had to run in and find some lee to get the water pumped out of her. That was a real dangerous situation."[28]

Perhaps no story epitomizes the essence of lake work in Fishtown more than this whimsical account of the *Good Will* from a 1950 issue of the *Leelanau Enterprise*. Titled "A Fish Boat Lonely as a Cloud," the front page article captures the danger, grit, ingenuity, and teamwork of fishermen pitting themselves against Mother Nature:

> This is the odyssey of the fish boat *Good Will* which for ten long days and nights has wandered off Leelanau shore on the wide cold waters of Lake Michigan. Locked out of her home port of Leland by a half-mile field of ice, the little boat with Lester Carlson and Louis Steffens aboard dodged storms, held on with toenails under the lee of North Manitou while a full gale roared from the north. She has fought zero temperatures and rain and fields of floating ice and the blindness of thick snow. The *Good Will* hurried out of Leland Harbor Sunday morning, March 12, with Gordon Carlson and Louis Steffens aboard. Ice was beginning to crowd the harbor, and the two fishermen decided to make a run for open lake. Twenty-two boxes of nets were missing on the banks off the Manitous. As the boat reached open lake a storm blew up. For several days the fishermen dodged around North Manitou, seeking shelter behind the island. They found five boxes of the nets. Then a slipping clutch caused trouble, next

Gordon Carlson smashed a finger. So the boat put in to South Manitou where the Coast Guard telephoned the mainland, and Lester Carlson went to Glen Haven to meet the boat. Ice at the dock prevented unloading the net boxes, but the Carlson brothers exchanged places, and the boat put out again. Friday a full gale with snow howled out of the north, and the boat huddled in the shelter of Lighthouse Point on North Manitou for a day and a night. Sunday she put into Frankfort where the net boxes finally came ashore. Refueled and restocked with food, out she pushed again. She still is out there, still hunting lost nets when weather permits.[29]

Henry Steffens dressing fish, c. 1930. PHOTO BY ERHARDT PETERS, COURTESY OF LEELANAU HISTORICAL SOCIETY

Chapter 9

Shanties and Shore Work

"The front part [of the shanty] was where they brought the fish in and weighed them and packed them up to ship them out. The back half of the shanty would have a net reel where they could work on their nets inside, when they came in off the lake. And these were all net reels out here [between the shanties]."[1]

– Bill Carlson, 2007

IMAGINE A BLUSTERY NOVEMBER AFTERNOON IN THE 1930S WHEN THE BIG Lake is too rough for fishing. Fisherman Henry Steffens sits in his shanty by the pot-bellied stove working on his nets. The smell of fish mingles with smoke from the stove. The exposed wall studs, joists, and ceiling rafters provide little protection from the cold wind that blows off the lake; the coal or wood-burning stove provides welcome warmth from the chill. There are few windows and two doors, one that opens to the river and the other to the alley that runs parallel to the river behind the shanties. The river side is considered the front door, and tradition dictates that Steffens will leave through this door at the end of the day. Every shanty has a bottle opener by the front door, for a Cola or something stronger. Steffens' fish tug, the *Helen S*, bobs at its moorings in the river outside.[2]

Like any working shanty, Steffens' contains the necessities of shore work: net reel, net and fish boxes, needles and twine for mending nets, corks and leads used in building nets, stove, molds for casting leads, a scale for weighing fish, buckets for mopping the floor, perhaps a buoy in need of painting or repair. A new litter of kittens mews in the corner, part of the cat population that protects stored nets from the gnawing teeth of mice. There is a countertop, sink, and carefully sharpened knives for gutting fish. A trap door in the floor serves as the on-site toilet. Other fishermen, especially the old-timers, drop by to pass the time.

Fish shanties were not all identical, and interior spaces changed to accommodate new technologies and increasing government regulation. Claudia Goudschaal remembered one day when Pete Carlson was "having a conniption fit, because the health department had been there, and they had said—decreed—that he had to put in a second sink, a separate sink for hand washing and a separate sink for fish washing. He was just really put out by going to the trouble and expense to do that. But he had to do it, so he did." Over time most fishermen modified the shanties, perhaps with new or larger windows to allow more light, or with additions to meet changing needs, such as space for ice machines, coolers, and retail counters. Especially on the river's south side where shanties were larger, fishermen made use of a loft created by covering the ceiling joists with planks.

Glenn Garthe shows the meter of a nylon net stored in the Leelanau Historical Society Shanty, 2010. In 1962 young Glenn worked for the Bucklers in this shanty. PHOTO BY LAURIE KAY SOMMERS

During Bill Carlson's boyhood of the 1940s, "I remember working upstairs in some of the buildings over there, because they'd build nets up there."[3]

The old-timers built nets to their specifications, stretching the meter with a winch along the full length of the shanty. (Meters are the heavier lines above or below a hanging net, to which the net is attached in order to strengthen it.) "Each net has two meters," Glenn Garthe explained. "On one meter you have corks tied to it. In the old days they had wooden corks, and then they changed over to the tied-on aluminum corks. Now the other meter was built with lead weights that were cast in Fishtown. We would melt the lead with blowtorches under a cast iron pot."[4]

After the fishermen stretched out the meter, Garthe recalled how "they would start slugging on twine, and tie on the corks, and hammer on the leads. Then they'd go to the next section." Fishermen were known to their fellow lakesmen by the distinctive characteristics of their nets. "Everybody had their own preference as to the sequence to tie a cork on and what kinds of knots they used," Garthe

Lead weights cast in Fishtown. PHOTO BY LAURIE KAY SOMMERS

Building nets, 1930s. PHOTO BY ERHARDT PETERS, COURTESY OF LEELANAU HISTORICAL SOCIETY

remembered. "We'd drag up somebody's net or find a chunk of nets along the beach, and the old fishermen like Pete Carlson or Roy Buckler could look at the cork tie and tell you where the net came from: which city in Wisconsin, or which city in Michigan."[5]

Fishermen spent much of their shore time tending their nets. "In the old days, the meter, twine, and the seaming twine was all cotton," Garthe explained, "and that's why the nets had to come out of the water and be dried." The labor-intensive cotton nets required periodic boiling in a large cast iron kettle or vat. "It was a lot of work back then," Bill Carlson recalled. "When they had cotton nets, they had to bring the nets in about every ten days or two weeks to dry them, scald them, treat them so they wouldn't rot. Now we can fish nylon nets, and we never have to bring them in except to repair them. They just don't deteriorate. They wear out after a while."[6]

During the 1950s, when Glenn Garthe began fishing, the fishermen were transitioning from cotton to nylon. "Obviously the mending of the nets, which we called slugging nets, had to be done on a regular basis," he recalled. "So somebody knew that this net needed repair—big holes in it. So it's brought to shore. You'd take the nets out of a box, and you'd rope them onto a reel and let them dry. I've done this outside, snowing and blowing. I remember Johnny Maleski being very upset with my uncles for making us go out and rope up nets in the middle of a blizzard." On days too stormy to fish, or during the wintertime when the boats were pulled up on shore, "you'd spend the day working on your nets," Garthe recalled. "Every shanty had a reel inside for slugging in winter."[7]

Net repair involved seaming on new twine or mending with needles. Filling needles was often a job for boys or for the old fishermen who liked to hang around the shanties. "There was always a box here, with rolls of twine on it," Glenn Garthe remembered. "You'd sit there in the chair and fill the box up with full needles. Then when you go to repair the nets, you always had full needles."[8]

Boiling cotton nets, here behind the Buckler shanty on the south side of the river, was an important part of shore work, 1930s. COURTESY OF JANICE FISHER

George Cook filling needles, 1930s. Piles of floats (also called corks) lie on the shanty floor. Floats were made out of wood, aluminum, and most recently, plastic. PHOTO BY ERHARDT PETERS, COURTESY OF LEELANAU HISTORICAL SOCIETY

The fishermen's first job after returning to shore was packing fish. "You didn't retail the fish," Bill Carlson explained, "You only needed an area to pack them." During the peak fishing period, only small numbers of fish were sold locally. The rest were gutted, iced and packed in fish boxes, and shipped by truck or rail to markets in Chicago, New York, and Detroit. "In the old days, when they'd pack fish, they didn't use cardboard like they do now," Glenn Garthe recalled. "We had wooden boxes that were built out of scrap wood. Emil LaBonte used to build them over here in Leland. We put one hundred pounds of fish in the box

Henry Steffens supervising his grandson, Glenn Garthe, at the net reel, 1961. COURTESY OF GLENN GARTHE

Gordon Carlson mending nets, 1940s. PHOTO BY ERHARDT PETERS, COURTESY OF DICK CARLSON

and covered it with ice. Every fisherman had their own tag, whatever fishery it was, and they would write on it, '100 pounds.' And that was stapled onto the end of the box."[9]

On occasion, if a customer wanted fresh fish, the shanties had counters or makeshift tables where fishermen would fillet, weigh, and wrap the fish. Herb Nelson remembered, "They had running water there, so they could wash the fish, and it drained into the river. All the innards, they would always carry down at

Marvin Cook dressing fish, early 1940s. The light-colored box in the lower right was a shipping box; the top slats extended for use as handles. PHOTO BY ERHARDT PETERS, COURTESY OF LEELANAU HISTORICAL SOCIETY

the end of the day to the end of the dock. The seagulls would always clean them up. Very sanitary!"[10]

Beginning in the late 1940s when the fishing declined, fishermen developed new moneymakers with the still plentiful chub harvest. Dick Carlson, son of Gordon, remembered making chub roe in the early 1950s. He and Leon Carlson—Pete's eldest son—would take the egg sacs, "squish them around a screen, and all the eggs would drop out. We would wash that to get all the blood out and lay the eggs on a screen to drain." The salted and dried roe was placed in wooden buckets and sent to Chicago or New York. "They'd make imitation caviar out of it," Dick Carlson explained. "We made some extra money that way."[11]

Chubs shipped to Chicago, New York, and Detroit also ended up as a smoked delicacy for the kosher market. The plentiful chub harvest yielded about half the price once received for lake trout, so to bolster profits fishermen began to smoke chubs for the small but growing local retail market. Smoked fish was not new to

Crew at Carlson's guts, fillets, and removes pin bones on the fillet table, preparing the daily catch for fresh market sale, 2010. Carlson's now trucks in fish from distant fisheries to supplement their catch, transported in vats of ice water to keep it fresh and in good condition. PHOTO BY LAURIE KAY SOMMERS

Fishtown. The earliest fish smoking was done by Ottawa and Ojibwa. European immigrants, especially those from Germany and Scandinavia, also brought their fish smoking techniques to the Great Lakes. Leland's smokehouses were initially the province of local grocers or butchers. During the 1920s fishermen set special pound nets for lake sturgeon that were sold to the Leland Mercantile, smoked, and then parceled out to local customers.[12]

The Carlsons were the first in the Leland area to smoke chubs. The *Grand Rapids Herald* reported in 1958, "They sell about five hundred pounds of smoked fish a week. Others in the area, fishermen and even store keepers, intrigued by smoked fish profits made by the Carlsons, soon followed suit until a brisk competition has been built up in the past ten years."[13] The Steffens family provided the competition in Fishtown.

Mike Grosvenor remembered the friendly rivalry between the Steffens and the Carlsons that occurred during his boyhood in the 1960s. "As a kid you'd

Henry J. Steffens Shanty and two side-by-side concrete block smokehouses, c. 1970s, with piles of maple for the smokehouse fire. The smokehouse to the left is no longer in existence; the one on the right still is used by Carlson's of Fishtown. Steffens' grandson, Leo Stallman Jr., was the last Steffens to smoke fish here, ending in 1973.
PHOTO BY GREG REISIG

Henry Steffens with his smoked chub hanging by their gills from a looped copper wire, 1967. COURTESY OF JOHN WESTOL

Alan Priest dips fish from the stainless steel brine tank (used today by Carlson's of Fishtown in lieu of the old ceramic crocks) and sorts them by type in the fish boxes, 2010. PHOTO BY LAURIE KAY SOMMERS

always go in and tell them that you like their fish best. And you'd play the game on the other side, too. They knew that you were just begging for a chub. We were like little vultures. We could time it when the chubs were coming out of the smokehouse nice and dripping hot. Pretty soon all the little urchins would show up for a handout. After we'd get our fill at Steffens, we'd go down to Carlsons and play the same game: 'Steffens just took their fish out of the smokehouse, but we would never eat *those* when yours were available!'"[14]

With the advent of fish smoking, shanties were outfitted with the necessary equipment such as brine crocks and fish smoking racks. The Carlson and Steffens families developed differing techniques for hanging the chubs in the smokehouse: Carlsons by the neck on nails, and Steffens by the gills on looped copper wire. They also made their own special hanging racks.

The Steffens grandchildren grew up helping their grandpa Henry Steffens

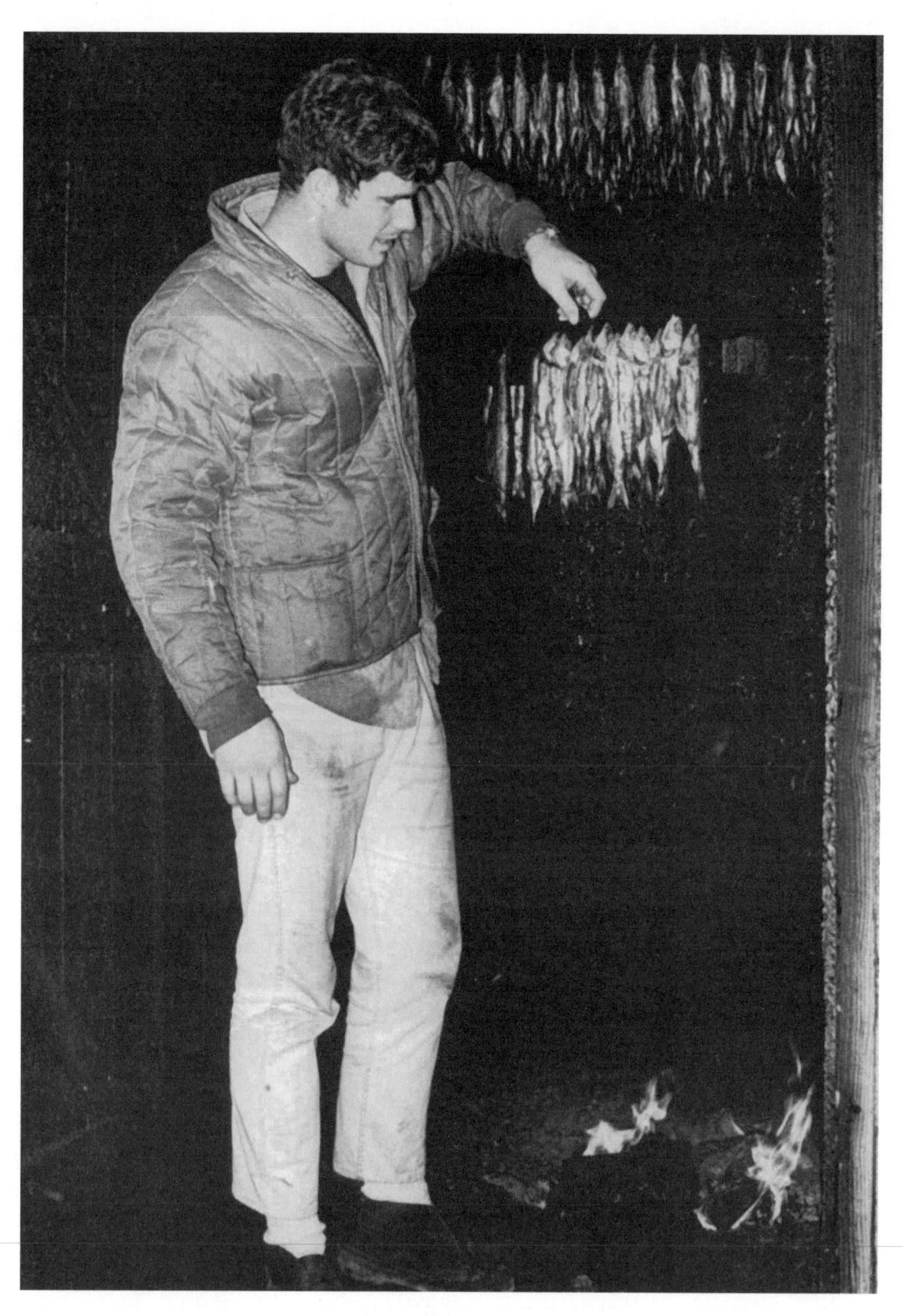

Bill Carlson checks a rack of chubs in the smokehouse, 1960s. COURTESY OF BILL CARLSON

smoke fish. "The old formula for making the brine was one pound of salt for every ten pounds of fish," Glenn Garthe recalled. "They used old ceramic crocks—big crocks—and they had homemade wooden hardwood lids. We would set stones on top to push the fish down under the water level. The next morning the fish were taken out—usually with a big wire mesh dip—washed, and placed on the smoking racks."[15]

The smokehouses "were all made of cinder block with a cement cap, as you can see today in Fishtown," explained Garthe. "You'd go through the process of lighting a fire like you would in a wood stove; it's paper and kindling. We used maple as the wood of choice." Today the entire process is heavily regulated, with temperature requirements, special probes, and thermometers. Fishermen originally had no modern technology to tell when the fish were done. "You looked at the color of their eyes and the color of the skin," Garthe explained. "I haven't smoked fish in years, but I remember if you burned a really hot fire fast, early in the stage of smoking them, then the bellies would open up. That's pretty much

Nels Carlson (left) and Alan Priest place fish by type on the smoking racks, 2010. Arranging the fish, Priest says, is "like a jigsaw puzzle." PHOTO BY LAURIE KAY SOMMERS

the way the Steffens did it."[16]

Before the botulism scare of 1963, "there was no shelf-life on the fish," recalled Glenn Garthe. "They were placed [in open boxes] and then stored in the cabinet in the shanty—no refrigeration. It had screen-covered doors, so there was ventilation, but the flies couldn't get to them." Henry Steffens peddled his fish locally into the 1970s.[17]

Carlson's of Fishtown still continues its family tradition of smoking fish. The expanded product line also includes smoked beef and turkey jerky, smoked trout sausage, and smoked whitefish pâté. In the height of the summer season, fragrant smoke billows from the smokehouse every day. Chub populations have declined dramatically, and the succulent smoked chub sell out quickly. The average customer joins a line at the retail counter or places an order ahead of time. For one local chub aficionado, however, there once was a special delivery method. Bill Hall, who bought the shanty across the river from Carlson's, would be hanging out with his buddies. "It would be cocktail hour," son Charlie Hall recalled, "and he would yell over to Bill Carlson, 'Why don't you airmail us some chub?' They would wrap a bundle—a pound, two pounds of chub—and chuck it across like a football."[18]

Chapter 10

Ice Houses
and the Community Ice Harvest

> "Leland's big annual winter 'bee,' the ice harvest, got under way yesterday morning and will continue until some time next week, depending on the weather."[1]
>
> – *Leelanau Enterprise*, 1931

"LAKE LEELANAU FROZEN NIGHT OF JANUARY 31." THIS WAS BIG NEWS IN 1931, when the *Leelanau Enterprise* reported that Mother Nature finally took pity on anxious "fishermen and others who need ice during the summer months. While it is not yet certain that the ice will become thick enough to be cut and stored before warm weather sets in again, the weather this week is favorable thus far. It is rumored, but no proof has been brought forward, that a quartet of fishermen sang 'Happy Days Are Here Again' last Friday night when the temperature began to drop. At any rate, they appear to be in better spirits this week than they have been in some time."[2]

In the age before coolers and ice machines, fishermen stored ice in the fishery's ice houses. The yearly "winter bee" was front page news not only for the fishermen's benefit. Merchants and households also needed ice, but commercial fishermen managed the ice harvest for the entire community. When an unseasonable warm spell delayed the usual starting date, everyone was concerned. The winter of 1931 was one of those years. "Until about a week ago, the ice on Lake Leelanau was hardly thick enough for the purpose," the *Leelanau Enterprise* reported. "Then, when it had attained a thickness of about ten inches, repeated thaws had left so little snow that sleighing [to transport the cut ice] was ruined. This caused another delay, pending further snowfall, but by Monday evening there was no more sleighing in Leland than there is in July."[3]

The ice harvest was a communal event that involved fishermen, merchants, and farmers. The men formed two crews, one to do the actual cutting and another to load the ice houses. The ice cutting crew usually worked in the area of south Lake Leelanau known locally as "the brown boat house" where it was easy to bring loading equipment and vehicles on and off the ice. Tasks included plowing and marking the ice, sawing and splitting the cakes, loading, and hauling.

Unloading boxes of fish cooled by shaved ice, c. 1930. PHOTO BY ERHARDT PETERS, COURTESY OF LEELANAU HISTORICAL SOCIETY

Working with the strips of cut ice in Lake Leelanau, 1930s. PHOTO BY ERHARDT PETERS, COURTESY OF LEELANAU HISTORICAL SOCIETY

Transferring ice cakes from sleigh to truck on Lake Leelanau, 1930s. PHOTO BY ERHARDT PETERS, COURTESY OF LEELANAU HISTORICAL SOCIETY

Edward Wichern's ice cutting machine, 1930s. PHOTO BY ERHARDT PETERS, COURTESY OF LEELANAU HISTORICAL SOCIETY

Oswald (Ozzie) Cordes—who with his brother Ralph owned the Leland Mercantile—described the ice harvests of his youth. "They had what was called an ice plough that scored ice with knives and would cut three or four inch furrows. Early in the morning someone'd go down with a crosscut saw made for cutting ice, saw a strip of it, and then break it loose with a spud." The loosened ice strip created a channel in the water that made it easier to saw and spud off individual cakes. "The man who was good at it could make a pretty square cut," Cordes recalled. Lifting the heavy cakes of ice required large ice tongs and a strong back. Protected by yellow oilskins and caulked boots, the men hoisted a cake and then dropped it back in the water. "The surge would help them lift it up on the ice," Cordes explained. "Then they would slide it onto a farm sleigh with teams of horses. Usually the farm people from East Leland would come over and haul the ice at so much per load."[4]

The ice crews continually sought ways to ease the work. In 1931 Edward Wichern developed a power-driven circular saw that cut almost through the ice. The *Leelanau Enterprise* reported, "It is then an easy matter for a man with a spud to split the cakes apart." This was not a job for the faint of heart. Eloise Telgard Fahs recalled that her father, Martin Telgard, accidentally sawed off a finger while cutting ice. Telgard, who Cordes described as "quite a man to do things in a mechanical way," developed an electric motor-driven conveyor for use on the ice.[5]

By Telgard's era fishermen used trucks to speed up the trips between Fishtown and the lake. The larger trucks carried thirty cakes per load and made four trips in an hour. In 1931 the *Leelanau Enterprise* reported, "But for motor trouble with Telgard's truck, which caused two hours delay, the all-time record would have been seriously threatened yesterday, for the amount of ice hauled in a day." The record number was never revealed, but it exceeded the 740 cakes the crew "brought to the north side of the 'fish creek' filling the ice house used by John Maleski and Kaapke & Firestone, and almost filling that of Henry Steffens and Harting & Light. That one was completed early this morning, and the trucks are now hauling for Oscar Price, and for William Buckler and George Cook on the south side of the river."[6]

Ice houses were simple wood-framed buildings typically located behind the fish shanties. Unlike shanties, they lacked conventional windows and doors but had openings at the eaves to provide light and ventilation, and access ladders next to slat-covered openings that extended from the ground to the dormered

The north elevation of Fishtown's ice house, c. 1930, then serving fishermen Steffens and Harting and Light. The building sports a white coat of paint, with the characteristic access ladder visible next to the shed dormer ice house opening. Vents in the eaves allowed light and ventilation (most visible in the far right building, the Maleski, Kaapke and Firestone ice house). A third ice house, partially visible to the far left with a sloping roofline, served Oscar Price. PHOTO BY ERHARDT PETERS, COURTESY OF LEELANAU HISTORICAL SOCIETY

The south elevation of Fishtown's remaining north-side Ice House, 2010. The two central doors replaced the original slat-covered opening once used to load and unload ice. The building was raised 3 feet in the 1980s to accommodate retail use on the lower level, so the ladder no longer reaches the ground. PHOTO BY LAURIE KAY SOMMERS

roofline. Many were reinforced inside with special sheathing that kept the walls from buckling under the pressure of packed ice and sawdust.[7] Although their height suggests a two-story structure, ice houses were built as one large interior space. As Fishtown expanded, so did the number of ice houses. By 1930 Fishtown had five ice houses. Today, two former ice houses survive: the Ice House, and the Warren Price Shanty, currently known as the Hall Shanty, which originally had an ice house in the east half of the building.

Initially, fishermen loaded the ice houses by hand, sliding the cakes with a two-runner slide into the ice house and stacking the cakes in layers separated by insulating sawdust from local sawmills. As Ozzie Cordes recalled, "They walked

The east half of the Warren Price Shanty, currently known as the Hall Shanty, served as an ice house. This photo shows the access ladder next to the dormer, 1920s. PHOTO COURTESY OF LEELANAU HISTORICAL SOCIETY

up and down all day long shoving." During Bill Carlson's boyhood in the 1940s and 1950s, ice trucks typically unloaded through the north door of the ice house, taking advantage of the higher grade of the road. The single "door" opening was made of removable boards. As stacks of ice rose higher, Carlson recalled, "they slid the boards in front of the door, all the way up to the top until it was filled with ice." Ladders next to the door enabled fishermen to reach the top.[8]

Daily ice kept the fish fresh. "You didn't store anything," Bill Carlson explained. "Nobody had coolers." Each day before heading out, the fishermen climbed the ladder along the south door of the ice house (visible yet today on the south façade of building). Working down from the top layer, they removed the necessary boards, dropped the day's supply of ice on the ground, and dragged the cakes to the river to rinse off the sawdust. Glenn Garthe is just old enough to remember the ice house era, which ended in Leland by 1960. The ice house sat just behind his grandpa Steffens' shanty. "They had a hand-operated machine for shaving ice blocks. We'd shave the ice right on the dock, scoop it up, and pack the fish. Then they bought ice machines; they had their own coolers, so then we had ice available." It didn't pay to be careless. Hands would get numb from the cold and a few fishermen nicked a finger or thumb along with the ice.[9]

New machines made the work easier. Bill Carlson recalled the first automatic shavers and the first automatic flaking machine. "And I remember when we

Loading a south-side icehouse with a runner slide, 1930s. PHOTO BY ERHARDT PETERS, COURTESY OF LEELANAU HISTORICAL SOCIETY

Percy Guthrie moving cakes of ice with ice tongs, 1930s. PHOTO BY ERHARDT PETERS, COURTESY OF LEELANAU HISTORICAL SOCIETY

Ice is still crucial to the fishery. Here Nels Carlson and Joe Campo of Carlson's of Fishtown unload boxes of fresh fish packed in crushed ice, 2010. PHOTO BY LAURIE KAY SOMMERS

got our first ice maker, which made big tubes of ice. We'd fill them with fresh water, let them set for a day, and then they'd make ice."[10] The ice harvest also mechanized over time, adding power saws and motorized conveyor belts, but it was still a dangerous and labor-intensive job.

Years after the ice house era ended, Roy Buckler reminisced with the *Leelanau Enterprise* about his experiences with the ice harvest. Buckler worked on a crew "of sixteen or more, two each from fishing boats in Leland. He spent about two weeks cutting ice for three large ice houses and three small ones in the Leland area [not all of these were Fishtown ice houses]. They would harvest about three thousand 'cakes' of ice and sell them for about three cents each. He was paid two twenty dollar gold certificates for about ten days work." After all their labors, fishermen celebrated the end of the ice harvest with big parties. Roy gave few details about the festivities. He did confide, however, "The wives didn't like it much."[11]

Chapter 11

Stories and Shenanigans

"The fishing season will soon be over but not the yarns."[1] – *Leelanau Enterprise*, 1904

FISHTOWN IS A PLACE OF STORIES. MIKE GROSVENOR REMEMBERED HOW HIS parents, George (Sonny) and Florabelle Grosvenor, often would go down to Fishtown before dinner "because there was always a gang of fishermen that were, of course, hard drinkers and storytellers. I remember as a little boy, the coal stoves going in the fish shanties, lots of tobacco smoke and laughter, and my mother wondering whether that was really a good setting for me to be growing up in. But my dad thought it was, so I usually got to go along."[2]

The old-time fishermen liked to hang around the shanty and tell tales. Alan Priest recalled how he used to listen to Roy Buckler tell stories about the lake for "hours and hours and hours. He always had a jug of vodka hid around here someplace in the shed. When we'd clean out and move boxes around—whoops, there's an empty jug there! Old Roy."[3]

Louis Steffens was a good storyteller. Nephew John Westol remembers sitting around the family dinner table as Louis would enthrall them with tales of being trapped in the ice. "He'd talk about the wind howling, the ice moving so you could hear it groaning, and how they'd radio for the Coast Guard to come."[4]

Although the Coast Guard provided assistance on countless occasions, fishermen loved to talk about out-toughing and outsmarting the Coast Guard. In the days when fish tugs worked deep into the winter, few other craft braved the frigid lake. Fishermen depended on their wits, their boats, and one another. "There's no help out there," Brian Price explained. "The story among fishermen was that the Coast Guard wasn't much help either. They would maybe respond if it was really a complete emergency situation. The fishermen felt they were more competent and their boats were more sound, especially in really bad weather. I've heard all kinds of stories about the Coast Guard guys leaving in the 40-foot boats and arriving somewhere half beat to death, half dead from exposure to the elements. One guy has a broken leg and another guy is half froze to death, and they're coming to help?"[5]

One Coast Guard tale involved what Mike Grosvenor called "the Fishtown secret"—the fishermen's ploy to "trick" the Coast Guard into freeing the harbor from winter's ice. In the early spring, when tantalizing open water appeared beyond the ice-clogged harbor entrance, "everyone was "antsy" to get out of port. They spent hours and hours battering the ice, trying to get to open water. The abuse they must have given their boats to get through that little bit of ice!" Once free of the harbor, there were no guarantees that the ice wouldn't re-form. Undaunted, the fishermen "would go out there and intentionally get stuck in the ice, to the point where there'd be two or three boats out there hopelessly caught in the ice. The ploy was to intentionally get stuck, then call the [buoy tender] *Sundew* in to break them out." The rescue, of course, involved breaking up the

Fish tugs trapped in ice floes, c. 1940. PHOTO BY ERHARDT PETERS, COURTESY OF LEELANAU HISTORICAL SOCIETY

harbor ice, so the fish boats could go their merry way the next day.[6]

Tourists were a frequent target for fishermen's shenanigans. "Marv [Cook] and Percy Guthrie really liked to tease the tourists," recalled Mike Grosvenor, "because this was an era where Fishtown was first starting to attract tourists as this quaint little fishing village. Marv would be out spinning nets up to dry, or mending them. Of course, that was a favorite picture of the tourists. When they'd start taking pictures of him, Marv would reach down, drop his pants, and just keep right on working with his bare butt sticking out under this shirt. This usually sent mothers and children scurrying up the hill. Eliminated some of the pictures, and probably attracted some other ones...."[7]

Fishermen also enjoyed poking fun at greenhorns. The Langs had a favorite story called "Brian and the Clinker." Clinkers, as described by Jim VerSnyder, are "an after-math of freighters burning coal. It is burnt slag that has a bunch of little spurs on it, and if gets in your net it is a real bastard to pick out." The Langs' story involved teasing their young crew member, Brian Price. "Brian was a college student," remembered Ross' wife, Joy. "He studied geology and thought he knew a lot about the lakes. One day they were pulling nets, and they got a clinker. So my father-in-law said to him, 'Well Brian, we don't know what this is. Could you tell us?' So Brian goes on and on with some long story about what this clinker was. Of course, the guys just had the best time over it because they knew it was a clinker."[8]

Marvin Cook, posing (modestly) with his net reels, 1930s. PHOTO COURTESY OF BARBARA GENTILE

Pete Carlson and Roy Buckler were Fishtown's consummate pranksters. Alan Priest recalled one incident when he was the butt of Pete's antics. "We come in the harbor during a big nor'wester, the first big sea I'd brought the boat in the harbor. You got a little tippy, but that's natural. Pete, he's waiting down there at the dock. I see him, and I thought, 'Oh, something's up.' He's chuckling, his shoulders are shaking, and he's got his hand in his coat pocket. When I get the boat docked and tied up, he brings out a roll of toilet paper and says, 'Here, thought maybe you might want to use this.' Then he started laughing. 'No, I'm all right,' I say. But he was just shaking like I had the s— stirred up."[9]

"They were a little different breed, no doubt about it," remembered Jim VerSnyder of the previous generation of fishermen. "They lived and breathed the water, and they fought it—they had to fight it—and they had fun with it when they wanted to. They were their own boss and free-spirited. And most of them liked to imbibe and tell stories." But the antics weren't confined to the old-timers. When questioned, Jim VerSnyder will reminisce about "some of the wild stuff we did at the fishery. After we'd get done fishing for the day, we'd staple a playing card to the fishery's wooden door. Then we'd start throwing knives and see who could stick that ace with a knife. We'd break tips off the knives. Of course, there was a little beer involved and maybe a shot of whiskey or two." VerSnyder laughs. "Then the next time we came down, there was a steel pipe coming down on the old sink. We were all drinking beer in glass bottles, and we decided we're going to do some bowling. If you hit the pipe it was a strike. So we're standing up by the display case rolling these beer bottles down there. And jeezus, there was a couple of cases of empty bottles all busted! Bill [Carlson] came down the next morning and he says, 'What did you guys do to the fishery?' And I says, 'You were right in the middle of that, my friend!'"[10]

During the seasonal smelt runs, when the river was black with fish, VerSnyder remembered how "we'd come in, in the middle of the day, and there'd be smelt just swirling in around here to eat. We'd throw our nets out and dip them and get a couple hundred pounds of smelt. You'd have two deep fryers going, have a half barrel of beer. Everybody in Fishtown would come in. We'd be cooking smelt, and everybody would be eating and drinking and raising heck. Smelt season was a really festive time down here with the commercial fishing operation. We always had a good time. We were just kind of a close knit group, and we spent a lot of time together. Even when we weren't working, we were still horsing around."[11]

The fishing is over for the season, but not the yarns…

Chapter 12

Saving Fishtown

> "I was saving something that was really important. Fishtown was very important. I still have dreams of when I was a child, about all of the activity down here—the people, the characters, and the family."[1]
>
> – Bill Carlson, 2007

CLAUDIA GOUDSCHAAL FIRST CAME TO FISHTOWN AS A TOURIST IN THE early 1950s. By that time several of the longtime fishermen had quit or died, leaving their shanties to the mercy of wind, water, and ice. "I had no idea what Fishtown was," she recalled. "Some of the fish shacks were sort of collapsing into the river. The shacks were unpainted with swayed roofs, and the sheds sort of leaned on each other. There was still one ice house standing on the south side. The dock, as I remember, was a little narrower than it is today with weathered old planks. When the [fisher] men were there, the shack doors at both ends were wide open for air circulation. People just walked right straight through, watched what they were doing, and walked right out on the other end. And there was always this wonderful fragrance of maple smoke in the air."[2]

Fishtown was undergoing a sea change, morphing from the vibrant fishing village of its prime into a struggling fishery better known as an artist colony and tourist destination. In 1939 Michigan State University had established its summer art school in Leland, and its students joined generations of artists who, since the late 1800s, had sought inspiration in its picturesque fishing village. By the 1950s artists were sketching buildings that bore the scars of decline and neglect.

A view showing the effects of neglect, early 1950s. The Price shanties, on the verge of collapse, were sliding toward the river. By the end of the decade the three shanties pictured on the river's south bank had been purchased by resorters. PHOTO BY WALTER McCORD

The surviving fishing operations barely had enough money to make ends meet.

Perhaps the most significant changes to Fishtown were invisible. Two of the buildings on the south side and one on the north were now owned by summer residents. Unlike Goudschaal, these were descendents of families who had vacationed in Leland for two or more generations. These resorters cherished an image of Leland as an idyllic summer playground, with a fishing village on the river. During the summer season, they bought fresh fish at Fishtown, hired the wives and children of fishermen (as well as other local families) as summer help, fished from the Fishtown docks, and—as youngsters—swam in the river, using fish tugs as diving platforms. The 2007 purchase of Fishtown buildings by the Fishtown Preservation Society (FPS) occurred in large part because of donations from summer resorters.

Leland's beginnings as a resort colony date to the turn of the twentieth century. Sale of the former iron company property in 1900 allowed the town to re-imagine itself as a summer retreat for well-to-do Midwesterners. As the lumber era ended, railroad companies that owned vast networks of former logging track began to promote northern Michigan as a resort destination. Chicago lumber interests behind the railroad expansion helped spread the word about the natural attractions of Leelanau County. Fountain Point Resort near the Leelanau Narrows opened its doors in 1889 to city dwellers seeking a summer retreat. During the 1890s the *Leelanau Enterprise* touted Carp Lake as having "the best fishing in the state," served by a popular fishing resort in the village of Leland where anglers could catch trout, whitefish, black bass, perch, and pickerel in a 'fisherman's paradise.'"[3]

Shortly after 1900, Leland's natural attractions drew families from Indiana and Iowa to its shores. Joseph Littell, a Presbyterian minister from Indianapolis, was among those who founded the resort colony of Indiana Woods between Lakes Leelanau and Michigan "in the unbroken, roadless, pathless forest south of the village." Littell later enthused, "The woods rang with the glad voices of children,

Art students sketch behind an abandoned Fishtown ice house, 1950. COURTESY OF SADA OMOTO

and the shores glowed with beach fires. Everyone was filled with delight and enthusiasm. Everybody told everybody else, and in this way Leland began its new day as a summer resort." Littell's list of early resort families included a cadre of successful Indiana professionals and businessmen: the Ball Brothers of Muncie, Indiana—Frank C., Edmund B., and Lucius—who became wealthy through the manufacture of Ball home canning jars; Ball in-laws Arthur Brady—a lawyer who by 1902 had become president of Union Traction Company of Indiana—and Hoosier Group artist J. Ottis Adams; Indianapolis attorney Frank Blackledge; Fort Wayne druggist August Detzer; and Ft. Wayne lawyer, superior court judge, and real estate man, H. W. Ninde. Around the same time, the Chandler family of Cedar Rapids developed another geographically organized colony, Iowa Shores, on the point in North Lake Leelanau south of the present-day Leland Country Club. "The initial cast of characters," wrote descendent Chandler Olds Higley, included Elmer Chandler, who platted Fishtown in 1908 and played a key role in the early development of Leland.[4] Resorters such as these helped to establish Leland as a playground for prominent and successful families, especially from the Midwest.

In the age before automobiles, vacationers traveled to Leland by rail or boat. While the women and children, typically with hired help, often stayed for the entire summer, the men could take the boat from Chicago, spend two days in the county with family and, after spending the night on a Northern Michigan Transportation Line steamer, be back at their desks on Monday morning. The *Leelanau Enterprise* reported "a great demand for summer cottages in Leland" and effused that "the summer of 1907 at Michigan's famed pleasure resort, Leland, promised to surpass the unprecedented one of last year. The hotels are open all the year around, visitors are always here in numbers, but the attendance is rapidly increasing and every day brings new guests."[5]

The "summer people" lived what long-time summer resident Caroline Brady called "parallel lives" with year-round Leland residents. Before the advent of the automobile, resorters rarely visited "downtown Leland" or Fishtown because of lack of transportation. Chandler Olds Higley recalled, "We children often walked to town or rode the Mercantile's delivery truck, but we had few good friends outside our bay, purely for lack of contact. In 1924 our family came by car for the first time and our tight little society on Iowa Bay would never be the same again." Later generations had more mobility, but differences in social class created a stratified society, as was the case in resort communities elsewhere.

Regular visits to Fishtown occurred most often by those summer people who lived nearby. For others, the summer season unfolded in a parallel world, with a bevy of luncheons, dinners, bridge games, and in later years, golf outings and cocktail parties.[6]

"Boats, real or toy," Chandler Olds Higley reminisced, "occupied much of our time." Before the advent of trailer hitches, most of the boating activity occurred on Lake Leelanau rather than Lake Michigan. "I would say the main activity growing up was sailing," recalled Elizabeth Wiese, a member of the Ball family. "There were races and regattas all the time on Lake Leelanau." An occasional yacht moored at Fishtown, either passing through or owned locally, and, as

Steamer approaching the Leland Dock Company dock, c. 1906. COURTESY OF BARBARA GENTILE

(ABOVE AND RIGHT) Commercial fisherman George Cook, friend to Barbara Gentile's family, taking the Scott boys fishing in his pound net boat, 1920s. COURTESY OF BARBARA GENTILE

remembered by Higley, "a few boats were brought up around the dam on rollers and launched in the Leland River for use in Lake Leelanau."[7] For the most part, however, the river below the dam was the domain of working boats until the 1950s, when resorters began renting slip space along the docks.

On the eve of the sea lamprey incursion, the Leland Township board listed the commercial fishing fleet as one of the attributes contributing to a "wholesome vacation" in its 1940 *Vacation Handbook*. Fishermen interacted with resorters and tourists from as early as 1900, when they used a fish boat or launch for excursion parties to North Manitou Island or Charlevoix. Fishermen befriended summer people and occasionally invited them on their boats. Barbara Gentile recalled fishing with Oscar and Vero Price on the *Wolverine* in 1933, the summer before it burned to the waterline and sank. "Oh we just went out to fish one day with them, my mother and brother and I. The *Wolverine* had an area in the front where you

could sit and fish just on the deck." Gentile also remembered playing "pirate ship" in the "big old pile driver they parked on the beach, and that was a lot of fun."[8]

More often, resorters visited with fishermen as they worked. Dick Ristine was part of a small group of summer youngsters who crewed on fish boats. The fishermen, he observed, "sort of tolerated the summer folks, but they also appreciated the fact that summer folks bought their fish, and they often rented their cottages, even their houses. I can think of two or three who moved under tents in the summer, and they rented the houses. Going down to Fishtown was an event each day," Ristine remembered. "I just loved to watch when the fishermen brought their loads in. Occasionally, back in 1920s, there were very, very heavy loads of fish. Captain Cook, I believe, had the record at that time. It was chiefly whitefish and trout."[9]

Rather than go out on the Big Lake, many local residents and summer people came down to the Fishtown docks to fish or watch the sunset, popular pastimes that continue. Later generations of children also swam in the river. "They'd dive off the dock and swim in the channel," Claudia Goudschaal remembered. "You kept your boat there. We rented space from Stallmans for our boat. And you had to be careful when you were starting up, because those kids would sometimes get behind the boat. They'd be standing on your propeller to see what was in the boat. Just having a good time. We'd chase them endless times, because you could do some serious damage if you didn't know there was a kid hanging off your propeller."[10]

As commercial fishing struggled in the 1950s, the resort community flourished. The *Detroit News* proclaimed, "Summer Colony in Leland Finds Life Easy and Gay," in a 1957 article that epitomized the economic and social divide between many local residents and summer people. "Local folks, discussing the resorters who occupy homes on Leland's two sapphire lakes, Michigan and Leelanau, say, 'Every summer one of the resorters gives a thousand dollar party sometime during the season.'" Mrs. Russell Riley, featured in the paper's story, replied, "'Only one party?'" Riley, the former Magene Mitchell, was the daughter of Detroit oil king Donald R. Mitchell, who first introduced his family to the beauties of Leland in

Fishing from the docks, c. 1917. COURTESY OF LEELANAU HISTORICAL SOCIETY

the 1920s. Although Mrs. Riley considered Leland "a down-to-earth summering spot," she admitted it still had lovely parties. "'But even these are purely sports clothes affairs, cocktail parties or barn dance,' Riley explained. 'I haven't worn an evening gown here since I was in my teens.'"[11]

Bill F. Hall and his wife, Sally, of Ft. Wayne were part of this social scene. A second-generation summer resorter whose father came to Leland in 1905, Hall became the first non-fisherman to own a Fishtown shanty. He bought the abandoned south-side shanty at the river's mouth from the Will Carlson estate in 1956. Hall's son, Charlie, was sixteen at the time of the purchase. "All I remember is, when I first saw the shanty it was just falling in the water, literally. The floor was out of it or falling into the water. The walls were wrecked, crooked. The board was basically deteriorated, an abandoned building that was used as an outhouse by people in Fishtown." As Charlie Hall put it, his father "just wanted a place to keep his tools, have his buddies down; it was a guy's place to go hang out." But the sociable Halls also participated in Leland's round of summer parties, and the shanty soon became a popular spot to entertain. Although the Halls did not acquire their shanty with preservation in mind, Bill Hall renovated the sagging structure and kept it from suffering the fate of the recently demolished Price shanties. The former Cook and Brown shanty, next door to Hall's new purchase, was in slightly better shape and still used by fishermen. The old Smith and Buckler shanty, two doors to the east, was sold in 1959 by fisherman Marvin Cook and his wife, Marie, to Adelia Ball Morris, daughter of E.B. and Bertha Ball, early residents of Indiana Woods.[12]

A long-time friend of Roy Buckler's, Morris understood that Fishtown's preservation depended on the continuation of commercial fishing in the Leland River. During her lifetime she rented the building to Buckler and his son Terry, and subsequently to Fred and Ross Lang. In 1985 she and daughter Barbara Goodbody donated the building to the Leelanau Historical Society with the proviso that it be used to support commercial fishing as long as it exists in Leland.[13] True to Morris' wish, the Leelanau Historical Society currently rents the shanty to Fishtown Preservation for storage, net repair, and to dock the *Janice Sue*, one of FPS's commercial fishing vessels.

Also in 1959, former fishermen Fred and Robert Buehrer sold their north-side shanty, originally built by Harting and Light. The purchaser was Carey Realty of Indianapolis, a family entity associated with summer resorter Alan Appel, partner in a successful insurance agency. Appel wanted dock space for his pleasure boat

Fisherman Alan Priest slugs nets in the Leelanau Historical Society Shanty, continuing its historic legacy of commercial fishing use, 2010. PHOTO BY AMANDA HOLMES

and had little interest in the shanty. Beginning in the early 1960s, he was the first to rent a shanty for retail use, initially to diver Jim Sawtelle and his wife Nan for their shipwreck furniture shop called Treasure Cove; then to Ken Krantz for his Inter-Arts gallery of Fishtown drawings; and next to summer resident Susann Craig for her gift shop, Limited, Ltd. Retail offered a low impact way to re-purpose abandoned shanties as the fishing economy continued to decline.[14]

Leland during the 1960s transformed from a summer retreat into a major tourist and pleasure boat destination. Fishtown underwent significant change as a result of this burgeoning tourist economy: completion of the new harbor of refuge and transient marina in 1969; advent of charter fishing and a revitalized sport fishery (late 1960s); creation of the Sleeping Bear Dunes National Lakeshore (1970); and the adaptive reuse of additional fish shanties for retail use. The Falling Waters Lodge and the Cove Restaurant, built on the east end of Fishtown in 1966 and 1967 respectively, also catered to the emergent tourist market.

The completion of Leland's long-awaited harbor of refuge coincided with the Department of Natural Resources' wildly successful salmon planting experiment; Pacific Coho and Chinook salmon bolstered the sport fishery and served as predators for the exploding alewife population. By 1966 invasive alewives made up 80 percent of the fish biomass in Lake Michigan and created havoc with ecological balance. The DNR also began to re-stock the depleted lake trout population as early as 1956, a decade before it planted salmon in various feeder streams (among them the Platte River that empties into Lake Michigan's Platte Bay between Frankfort and Leland). Long-time Leland charter fisherman Jim Munoz remembered how the sport fishing era began. "In '68 when I moved up here, there was very little sport fishing before that time. By '67 was the start of it on Lake Michigan with the advent of the salmon and that first salmon run. My

A massive die-off of alewives choked the river entrance during construction of the Leland Harbor of Refuge, c. 1967.
PHOTO BY MIKE BROWN

God, you would not believe what people did that first year. It was unreal. There were thousands of boats on Platte Bay. Guys would go out in a bathtub if they could. They would go out in anything, and these huge big fish were everywhere."[15]

Prior to 1972 Leland had one part-time charter boat captained by Floyd Lawson. In that year Jim Munoz and Jack Duffy both got licenses, although Duffy did not move to Leland until 1978. "This season will be our thirty-sixth year," Duffy recalled in 2008. "We were the only two boats here, and we had a heck of a fishery and just tons of lake trout. The fishing was fantastic, and word got out all over. Every year it seemed like somebody new would come in. Probably around the early 1980s, we had eighteen charter boats."[16]

The new harbor and marina drove Fishtown's thriving tourist economy. "I think a lot of people underestimate how much revenue that harbor brings in," Bill Carlson observed. "I think probably as far as the economic impact on Leland, and what makes things roll, the harbor probably has as much of an impact as anything."[17] Fishermen rode the tourism wave as it washed over Leland. Hank

John and Jeff Thompson posing with salmon after a successful charter trip with Captain Jack Duffy on the *Whitecap*, 1979. The fish tug *Mary Ann* is in the background. PHOTO BY JOAN FISHER WOODS

The Ice House, pictured here in 1981, was adapted to retail use in 1970. PHOTO BY LAWRENCE M. SOMMERS

Steffens and Leo Stallman Sr. had been forced out of fishing in 1968, but they and the elder Henry Steffens continued their fish smoking businesses into the 1970s. Steffens and Stallman rented their other former fishing buildings for tourist shops. Ken Krantz remodeled the old Steffens net shed for his Inter-Arts store. In 1969 summer resident Richard Braund built for Steffens and Stallman a "new shanty" intended for retail use where Braund opened his Reflections Art Gallery. In 1970 third-generation summer resident Lana Gits of Chicago renovated the ice house that became obsolete after the 1960s when fishermen adopted electric freezers. Gits was just eighteen when she began renting the ice house for a shop that featured arts and crafts, baked goods, and fudge.

The success of these initial shops helped solidify Fishtown's reputation as a tourist destination. Additional impetus came with the establishment of the Sleeping Bear Dunes National Lakeshore in 1970, and most importantly, the inclusion of South Manitou Island by the early 1970s and North Manitou Island in 1984. Manitou Transit, operated by the Grosvenors from Fishtown, first received the National Park Service concession for ferry service to the islands in 1986. The federal government created the Manitou Passage Underwater Preserve

Falling Waters Lodge under construction, 1966. The Suttons Bay architectural firm of Arai and Hummel designed the mid-twentieth century modern building for clients Fred and Noreen Hollinger. Its construction, along with a sister building, Fisherman's Cove, spurred Bill Carlson's initial efforts to preserve Fishtown through purchase of properties that came up for sale. PHOTO BY MIKE BROWN

in 1988, which became a tourist attraction for diving enthusiasts. The emerging tourism economy also included the construction of Falling Waters Lodge and Fisherman's Cove Restaurant in the mid-1960s. The massive scale of the two concrete buildings, with their wood shingle-clad mansard roofs, dwarfed the smaller wooden buildings downstream. They created the most dramatic change of character in the history of Fishtown. For Bill Carlson, the two buildings "changed the authenticity of Fishtown." He had a different vision, grounded in his experience as a fourth-generation fisherman.[18]

Instead of building new tourist attractions, Bill wanted to expand the retail use of existing shanties while keeping the site layout true to its past. He and his brothers began purchasing former fishing buildings as they became available, beginning in 1977 with the Steffens and Stallman property on the river's north bank and the former Cook and Brown shanty to the south. Bill also developed

Bill Carlson built a two-story retail shop on the northeast corner of Fishtown, pictured here in 2010, to replicate the look of the old ice houses. The former Louis Steffens Shanty (on right) has been the site of a cheese business since 1980. PHOTO BY LAURIE KAY SOMMERS

his own retail businesses to occupy the buildings and keep the tourist business growing. "Having more retail space was fine," he explained. "We could've put in a lot more, but [it] wouldn't have been authentic." Carlson's goal was to remain true to the Fishtown of his boyhood.[19]

The last property Carlson purchased, the old Harting and Light shanty, belonged to the Appel family. By this time Carlson Properties was in debt and seeking another entity to take over Fishtown. "We founded the Fishtown Preservation Society to try to keep us in the fishing business," Bill Carlson recalled. "We set this all up in 2001, saying 'Fishing is my life, not Fishtown. If I can't fish, I'm going to sell Fishtown.' That got things started." The mission of Fishtown Preservation soon shifted to saving Fishtown as an active fishery and to preserving the historic integrity of its buildings. From the beginning, Carlson hoped he would find a buyer who shared his vision. "It was important for us that it stay this way," he reflected, "but sometimes it's not in your power once you lose control of it."[20] Fortunately, in 2004, the Carlsons turned FPS over to a group of dedicated citizens who sought to purchase and preserve the property as both a historic and commercial attraction. By 2006 FPS had reached an agreement to purchase the Carlson family's Fishtown interests. Within seven months—in an astoundingly successful flurry of fundraising—the organization and its "Save Fishtown" campaign raised $2.7 million towards this goal. Finally, on 7 February 2007, FPS completed acquisition of the land and structures between the Cove Restaurant and Manitou Transit on the north side of the Leland River, the two fish tugs *Joy* and *Janice Sue*, and their equipment and licenses. Fishtown was saved, but mortgage payments and the challenging work to maintain and interpret a historic working waterfront was just beginning.

This triumphant moment was a testament to countless individuals who worked to sustain Fishtown through good times and bad: fishermen and their families, summer residents, tourists, boat builders, carpenters, mechanics, local community members, artists, merchants, and everyone who told its stories. People from all walks of life joined together to protect a cherished place. Few individuals understand Fishtown's special qualities better than the fishermen. Alan Priest works at Carlson's Fishery and captains the *Janice Sue*. "I grew up down in Fishtown," he explains. "When I was a kid, I fished off the boats. Then I grew up to fish on the boats. That's my second home down there. It needs to be taken care of."[21]

Thankfully, Fishtown Preservation is doing just that.

Afterword

Fishtown and the Future

by Kathryn Bishop Eckert (Omoto)
Chair of the Board of Directors, Fishtown Preservation

WE'LL BE HERE NEXT YEAR.

That may be the simplest way to explain what Fishtown Preservation is all about. When everything else in our lives changes, we want to know that Fishtown will still be here for us, next year. However many different things Fishtown means to each of us, it's what brings us all together in this place.

Americans have always loved progress and the new. Our search for better lives for ourselves and our families means constant change, but one unexpected consequence is that few things last in this country, even those that we treasure.

Fishtown has long been beloved by Leland locals and cottagers as well as visitors, but that affection has not shielded it from the effects of change. "Its unique collection of fishing shanties is unparalleled on the Great Lakes," wrote one observer. "Unfortunately, the trend is for commercial fishing to disappear from Leland. The outlook even for next year is uncertain. When this happens, much of the charm and character of Fishtown will have disappeared as well."

These words of warning came not during the recent "Save Fishtown" campaign, but in 1975. The writer, Alan William Moore, had spent the previous summer on a research project for Michigan Sea Grant, looking for a way to preserve Michigan's heritage of commercial fishing. Leland's Fishtown was one of five places he visited as possible sites to preserve a historic fishery.

What Alan found here was familiar: a place like no other (even then), with many shanties in deteriorating condition, without enough funds to maintain them, amid an industry under grave threat. Fortunately, he also saw glimpses of why Fishtown remains alive today: not only those unmatched shanties, but also local residents interested in preservation and fishermen conscious of their own past.

In fact, I was one of those who helped point Alan toward Fishtown. To identify historic fisheries and discuss preservation tools and ideas applicable to the fishing industry, he came to Michigan's State Historic Preservation Office, where I'd just begun working. The office had opened a few years earlier, in 1969, to carry out our state's part of the National Historic Preservation Act of 1966. That law had passed in response to public alarm about how much of the fabric of our country we were losing to projects like urban renewal and the interstate highway system.

Fishtown had recently been listed on state and national historical registers—a promising sign—and Alan ended his Fishtown report on a hopeful note. "A modern commercial fishing operation should be maintained at Leland at all costs," he wrote. "A concerted effort by local citizens might accomplish this goal."

In the decades that have passed since Alan's report, little happened to preserve the historic fisheries he'd identified—except in Leland, where "a concerted effort by local citizens" actually did save Fishtown, not as a modern commercial operation, but as an historic site. The properties are now in the hands of the Fishtown Preservation Society, Inc., a non-profit Section 501(c)(3) corporation, that acquired them from the Carlson interests in 2007. As a result, Fishtown was spared from the commercial development that obliterated so many historic areas up and down the Michigan shoreline. Today, Carlson's Fishery continues to market fish, both wholesale and retail; the shanties that create the Fishtown experience have been stabilized and plans are underway for their maintenance and rehabilitation; and two historic Fishtown tugs, *Joy* and *Janice Sue*, still fish Lake Michigan.

All this was made possible by the generous contribution of $2.7 million by hundreds of people who understood the value of Fishtown and its importance to the Leland community. The Fishtown Preservation Society and its donors saved Fishtown. But "saving" a place as dynamic, meaningful, and vulnerable as Fishtown is an ongoing process.

I've devoted my career to helping Michigan communities identify, review, and protect properties worthy of preservation. I've always enjoyed learning and sharing everything I could about our state's architecture. Even more, I've loved helping people to protect, enhance, and enjoy the places that matter most to them.

I am part of Fishtown Preservation today because Fishtown matters to me. You

The Fishtown Preservation Society's 2007 purchase includes eight shanties on the north side of the river. PHOTO BY LAURIE KAY SOMMERS

might not realize it, but Fishtown also matters on a much larger scale. Fishtown aligns our local concerns with broader preservation issues nationwide. In preserving a real, historic and working Great Lakes waterfront, we are demonstrating the power of preservation to contribute to our community and our region.

Historic preservation in America is undergoing a generational transformation. Following the crises of the 1960s, support among Americans for preserving our heritage has grown, as have preservation organizations. Now we must deepen this foundation, to demonstrate in concrete terms that preservation will support not only what we value from the past, but our prospects for the future as well.

Fishtown Preservation is a success story. We face many challenges, of course, and sometimes feel a little daunted at how much remains to be done. Still, we are systematically following through with long-range projects, each one building upon the one before. We are showing the kind of progress that leads to long-term success.

We completed our Master Plan in early 2009 to provide the high-level road map that would allow us to focus our preservation efforts. In 2011 we took the next step, completing the first definitive account of Fishtown in *The River Runs Through It, A Report on Historic Structures and Site Design in the Fishtown Cultural Landscape* (HSR). Combining the efforts of historians, folklorists and building and landscape architects, the HSR brings together records and stories of the people. It also documents and assesses the physical condition of the structures and site—including a state-of-the-art, high-definition 3D spatial scan of the entire Fishtown area—and then presents a preservation treatment approach. The report will ensure that all future work—stabilization, repair, rehabilitation, and maintenance—will respect Fishtown's history and significance, existing conditions, and ongoing use as a fishery. Intended as an internal guide for our work, the HSR is drawing reactions of astonishment that tell us that it is something much more significant. We will pass on its information in as many forms and forums as possible.

Much has changed in Fishtown since Alan's 1975 report. Fishtown's story, however, is one of resilience; change is not the same thing as loss. Many of the faces in Fishtown have changed, but much of the way and pace of life remain. Even the Falling Waters Lodge and Cove Restaurant, whose mid-1960s addition to the Fishtown environment was then so controversial, we can now appreciate as examples of mid-twentieth century modern architecture. Change is inevitable. What Americans have learned over the past several generations is that, instead of being victims of change, we can help to shape the course of progress.

Looking back, I can see that all I have done has led me here, to Fishtown. Art, architecture, historic preservation, issues national and local—Fishtown is where the puzzle pieces all fit together. I take great pleasure in working to make Fishtown a place that lasts—because Fishtown is a place that matters. I embrace Fishtown Preservation because what we value now will be our legacy to the future.

That's what I mean when I say: *We'll be here next year.*

Acknowledgements

The Fishtown Preservation Society would like to thank the many generous donors who have made possible this and other recent projects exploring and enhancing our understanding of Fishtown's rich history. The book was developed out of the extensive research completed as part of the 2011 Fishtown Historic Structure Report (HSR), *The River Runs Through It: Report on Historic Structures and Site Design in the Fishtown Cultural Landscape*. In support of that project, the Midwest Office of the National Trust for Historic Preservation awarded FPS a grant from the Jeffris Heartland Fund, which was made possible by the Jeffris Family Foundation. That seed money set the stage for other generous grantors. The Edmund F. and Virginia B. Ball Foundation, the Dogwood Foundation, the Americana Foundation, Alexander and Sally Bracken Family, Jennie Berkson and David Edelstein, Carol F. Maxon, and Edward and Lisa Neil all stepped forward to help match the grant. Digital scans of Fishtown, key to the presentation of the site and building studies, were funded through a grant from the Michigan Coastal Zone Management Program, Department of Environmental Quality. An extensive collection of oral histories was supported by a National Oceanic and Atmospheric Agency Preserve America Initiative Grant, part of Preserve America, a White House program aimed at preserving, protecting, and promoting our nation's rich heritage.

The publication of *Fishtown* is made possible by a grant from the Michigan Humanities Council, an affiliate of the National Endowment for the Humanities. Any views, findings, conclusions or recommendations expressed in this project do not necessarily represent those of the National Endowment for the Humanities or Michigan Humanities Council. The Les and Anne Biederman Foundation and an anonymous family foundation gave generously in support of this matching grant.

We are grateful to the dedication of the Fishtown Preservation Society Board of Directors that supported the planning projects that have led to this history of Fishtown and its fishermen. The directors, in alphabetical order, include Keith Ashley, Sandra Clark, Berkley W. Duck III, Kathryn Bishop Eckert (Omoto), Fred Heslop, Wilfred J. Larson, Dan McDavid, Craig A. Miller, James D. Ristine, Paul Sehnert, Cris Telgard, Ann K. Watkins, and Joan Fisher Woods. Their vision guides the future care of Fishtown and has made it possible for us to share Fishtown's history with the public.

Throughout the project we were assisted by many able individuals and organizations. Brandon Walker, PE, of Midwestern Consulting, LLC, conducted the laser scanning of Fishtown and provided the scanning images used as the basis of the site maps. Staff at the following institutions assisted with research: Archives and Library of Michigan, Lansing; James McClurken Private Library, Lansing; Northwestern Michigan College Osterlin Library and Archives, Traverse City; Traverse Area District Library, Traverse City; Traverse Area Historical Society, Traverse City; Leelanau Historical Society, Leland; Michigan Maritime Museum and Great Lakes Research Library, South Haven; Bentley Historical Library, University of Michigan, Ann Arbor; Leelanau County Treasurer's Office and Register of Deeds Office, Sutton's Bay; and Leland Township Assessor's Office, Lake Leelanau.

Special thanks go to James M. McClurken for providing access to his outstanding private collection of American Indian materials, and to Judith Schlaack and Michael J. Chiarappa for making available interviews conducted for the Fish for All Oral Histories (1999), a project of Western Michigan University and the Great Lakes Center for Maritime Studies. For helping us to meet a compressed publishing deadline, we'd like to thank Marcy Branski, for her eye for detail and attention to language; and Daniel Stewart of History by Design, for bringing together all the elements into the design of this book.

Interviews, including the Fish for All group, enriched this work immeasurably. Many additional interviews had been completed when the HSR study began and have been used in developing this work: in the 1980s and 1990s by members of the Leelanau Historical Society; and since 2007 by Amanda Holmes, Daniel Stewart, and Teresa Scollon for FPS. Many individuals shared recollections both formally—in recorded interviews—and informally through conversations by phone, email, or in person: Phil Anderson, Dan Appel, Caroline Brady, Richard Braund, Mike Brown, Bill Carlson, Clay Carlson, Dick Carlson, Mark Carlson, Nels Carlson, Susann Craig, Jack Duffy, Kathryn Eckert, Betty Elliott, Eloise Fahs, Glenn Garthe, Barbara Gentile, Lana Gits, Barbara Goodbody, Claudia Goudschaal, David Grath, Mike and Beth Grosvenor, Charlie and Mike Hall, David Hanawalt, Roger A. Hummel, Ken Krantz, Joy Lang (Anderson), Nick and Susann Lederle, Phyllis LeMar, Elizabeth Marshall, Jim Munoz, Herb Nelson, Sada Omoto, Ed Peplinski, Brian Price, Bruce Price, Alan Priest, Bill Rastetter, Dick Ristine, Don Strayer, Marilyn Stallman, Leo Stallman Jr., Ann Steffens, Cris Telgard, Jim VerSnyder, John Westol, Elizabeth Wiese, and Joan Fisher Woods. Bill Carlson, Alan Priest, and Jim VerSnyder also assisted in preparation of the traditional fishing grounds map.

Numerous individuals have shared photos and scrapbooks with FPS that are incorporated into this book. We are especially appreciative of the Bluebird Restaurant, which graciously allowed us to use photographs from its Fishtown photo collection, Barbara Gentile and Sada Omoto who shared scrapbooks, Meggen Watt Petersen who documented the fishermen at work, Mike Brown who made available his collection of photos and other Fishtown memorabilia, and the following organizations and individuals: the Leelanau Historical Society, Keith Burnham, Janice Sue Kiessel, Greg Reisig, Janice Fisher, Joan Fisher Woods, Glenn Garthe, Tom Kelly, Walter McCord, Joy Lang Anderson, John Westol, and Bill Carlson. Several firms generously assisted with graphics and maps: Cools and Courier, Inc. of Okemos; HopkinsBurns Design Studio of Ann Arbor; Johnson Hill Land Ethics of Ann Arbor, and Michigan Sea Grant. Michigan State University Press granted permission to reprint a map from the 1977 *Atlas of Michigan*. Finally, Bill Carlson has patiently answered emails, phone calls, and repeated again and again "who built what when." We hope this book does justice to all the wonderful information and assistance we have received.

Notes

A Place Called Fishtown

[1] The *Leelanau Enterprise* referred to the area as "fish town" in a 27 July 1939 article. This is the earliest known use of the name "Fishtown."

[2] The 29 February 1931 article of the *Leelanau Enterprise*, "Leland Puts Up Ice For Summer, Annual Harvest Begun Yesterday, Later Than Usual," refers to the north side of the "fish creek." Various conversations and recorded interviews attest to the term's continuing usage, among them Mike Brown's discussions with Laurie Kay Sommers on 22 August 2010, and Dick Carlson, in his 30 November 2010 interview for the Fishtown Preservation Society (FPS), recorded by Sommers in Grand Rapids, Michigan.

[3] In 2010 the *Joy* met her fishing quota, but the *Janice Sue* was pulled early as she only lifted 1,400 of a possible 90,000 pounds of chubs, too few to warrant the expense of continuing to fish.

[4] Alan Priest, interviewed by Daniel Stewart for FPS, 27 February 2009, Cedar, Michigan.

Why Fishtown Matters

[1] Alan W. Moore, "Michigan Historical Fisheries: Policy and Preservation Considerations Involved in Their Establishment in a Michigan Shoreline Community" (M.S. thesis, University of Michigan, 1975), 107.

[2] Dan Egan, "Commercial fishing, once a Great Lakes way of life, slips away," *Los Angeles Times*, 30 August 2011, http://articles.latimes.com/2011/aug/30/business/la-fi-great-lakes-fishing-20110830; Dustin Dwyer, "The Shrinking Commercial Fishing Industry," *The Environment Report: Swimming Upstream*, 23 June 2011, Michigan Radio, University of Michigan, Ann Arbor, http://environmentreport.org/swimming_upstream.php; Mary Russell, "Commercial Fishing Was Once a 'Major Industry' for Some Communities."

Preview Community Weekly, 21 May 1990; Joel Petersen quoted in *FPS Newsletter*, Summer 2011, 9.

[3] Amanda Holmes, quoted in Kristine Morris, "The Invisible Harvest," *Grand Traverse Insider*, 10 January 2011, http://www.morningstarpublishing.com/articles/2011/01/10/grand_traverse_insider/news/leelanau_area/doc4d2b5ef854c94991845184.txt?viewmode=2; Amanda Holmes, "When Pentwater Had a Fishtown," FPS, 6 May 2011, http://fishtownmi.org/2011/05/pentwater-fishtown/.

[4] Bill Carlson, interviewed by Amanda Holmes for FPS, 9 August 2007, Leland, Michigan; Jan Weist, "A Model Leland Fisherman Casts Lot with Tradition," *Detroit News*, 24 October 1977.

[5] Amanda Holmes, "Who We Are," Fishtown Preservation, www.fishtownmi.org.

[6] Amanda Holmes, "Help with description of other surviving fishery buildings," email to Laurie Kay Sommers, 4 April 2012; Holmes, quoted in Morris, "The Invisible Harvest"; "A Great Lakes Fisherman Learns Ancestral Lesson," *Christian Science Monitor*, 6 September 1994.

[7] "Michigan Leads in Great Lakes Fishing Industry," *Leelanau Enterprise*, 27 March 1930; "Leland Harbor, Michigan," Letter from Chief Engineers, United States Army, Report of the Board of Engineers for Rivers and Harbors on Review of Reports Heretofore Submitted on Leland Harbor, Mich., With Illustrations, 17 May 1935.

[8] Bill Carlson, interview, 9 August 2007.

[9] Fishing license numbers provided by Michigan Department of Natural Resources Fisheries Division Biologists Thomas Goniea, phone conversation with Laurie Kay Sommers, 26 March 2012, and David Carrofino, email to Laurie Kay Sommers, 28 March 2012; Kimberly Hirai, "Commercial fishing operators, scientists study Great Lakes whitefish fluctuations." *Great Lakes Echo*, 15 April 2010, http://greatlakesecho.org/2010/04/15/commercial-fishing-operators-scientists-study-great-lakes-whitefish-fluctuations/.

[10] Kenn Haven, "Harbor Won't Interfere, Leland's Fishtown is Saved," *Detroit Free Press*, 27 July 1966.

[11] Leelanau County Economic Development Committee, *Leelanau County "Land of Delight" Overall Economic Development Program for Leelanau County, State of Michigan*, submitted to the Economic Development Administration (1966), 52, 58.

[12] Kathryn Eckert, email to Laurie Kay Sommers, 1 April 2012; Moore, "Michigan Historic Fisheries," 104,107.

[13] Ned Kaufman, *Place, Race, and Story* (New York and London: Routledge, 2009), 38.

[14] Jim VerSnyder, interviewed by Daniel Stewart for FPS, 12 May 2008, Leland, Michigan; Dick Carlson, phone conversation with Laurie Kay Sommers, 29 March 2012; Bill Carlson, email correspondence to Daniel Stewart, 15 June 2011; *FPS Newsletter*, Summer 2011, 1-2.

[15] *FPS Newsletter*, Summer 2011, 1-2.

Before Fishtown

[1] *Leelanau Enterprise*, 15 September 1927.

[2] Alexander Winchell, *A Report on the Geological and Industrial Resources of the Counties of Antrim, Grand Traverse, Benzie and Leelanaw in the Lower Peninsula of Michigan* (Ann Arbor: Dr. Chase's Steam Printing House, 1866), 10, 12.

[3] Bela Hubbard Papers, 1814-1896, Box 1, Survey Notebook 2, Field Notebooks May 19-July 24, 1838, Peninsula Coast Survey, Detroit to Chicago, Bentley Historical Library, University of Michigan.

[4] James McClurken, Affidavit, Grand Traverse Band of Ottawa and Chippewa Indians v. Director, Michigan Department of Natural Resources, Leland Township, and Village of Northport, United States District Court for the Western District of Michigan, 14 April 1995, 6; Variously spelled as Shemagobing or Che-ma-gobing; McClurken, Affidavit, 5.

[5] H. R. Page and Co., *The Traverse Region Historical and Descriptive, and Portraits and Biographical Sketches* (Chicago: H.R. Page and Co., 1884), 242.

[6] *Leelanau Enterprise*, 15 September 1927.

[7] *Michigan State Gazetteer and Business Directory*, 1867; H. R. Page and Co., *The Traverse Region*, 243-244; Joseph Littell, *Leland, An Historical Sketch* (Indianapolis Printing Company, 1920), 29-30; Edmund M. Littell, *100 Years in Leelanau* (Leland: J. William Gorski Historical and Genealogical Collection, 1965), 36.

[8] Frederick W. Dickinson, *A Short History of the Leland Iron Works*, annotated by Harley W. Rhodenhamel (Leland: Leelanau Historical Society), 1996.

[9] *Leelanau Enterprise*, 20 November 1884, reprinted in *Leelanau Enterprise Special History Edition*, 14 December 1944.

[10] Letter to the Honorable Arthur H. Vandenberg from James W. Good, Secretary of War, 20 September 1929, containing the Report on Leland Harbor, Lake Michigan, by the District Engineer in Milwaukee, Mike Brown Collection, Leland, Michigan; *Leelanau Enterprise*, 1 March 1900.

[11] Karl Detzer, "Happy 100th Birthday Leelanau," *Leelanau Enterprise*, Centennial Edition, 25 May 1963.

[12] Littell, *Leland, An Historical Sketch*, 32.

Seasons of a Fishery

[1] *Leelanau Enterprise*, 25 October 1938.

[2] Littell, *Leland, An Historical Sketch*, 32.

[3] Michael J. Chiarappa, "Great Lakes Commercial Fishing Architecture: The Endurance and Transformation of a Region's Landscape/Waterscape," in Kenneth A. Breisch and Alison K. Hoagland, eds., *Perspectives in Vernacular Architecture: Building Environments* 10 (2005), 217-232.

[4] Pete Carlson and Roy Buckler, interviewed by Leelanau Historical Society, 1987, Leland, Michigan.

[5] Moore, "Michigan Historic Fisheries"; Margaret Beattie Bogue, *Fishing the Great Lakes: An Environmental History, 1783-1933* (Madison: University of Wisconsin Press, 2000).

[6] Roy Buckler, interviewed by Laura Quackenbush for Leelanau Historical Society, 1991,

Leland, Michigan.

[7] Carlson and Buckler, interview, 1987; Carlson quoted in Jim VerSnyder, interviewed by Laurie Kay Sommers for FPS, 27 September 2010, Leland, Michigan.

[8] Carlson and Buckler, interview, 1987; Buckler interview, 1991.

[9] Buckler, interview, 1991.

[10] Bruce Price, interviewed by Laurie Kay Sommers for FPS, 19 August 2010, Lake Leelanau, Michigan.

[11] Ibid.

[12] Russell, "Commercial Fishing."

[13] Carlson and Buckler, interview, 1987.

[14] Carlson and Buckler, interview, 1987; Leland Historic District, National Register working files, Michigan State Housing and Development Authority, State Historic Preservation Office, Lansing, Michigan; Carlson and Buckler, interview, 1987.

[15] "Fisherman 'Calls it a Day.'" Newspaper article, Janice Sue Kiessel Collection, FPS [1954?]; *Leelanau Enterprise*, 7 August 1947; Bill Carlson, interviewed by Michael J. Chiarappa for Fish For All Project, courtesy of Michael J. Chiarappa and the Great Lakes Research Library, South Haven, Michigan, 26 and 27 May 1999, Leland, Michigan.

[16] "Fisherman 'Calls it a Day;'" Amanda J. Holmes and Daniel W. Stewart, "We're All Brothers Out There" Voices from the Leland Fishery at Fishtown, 1940-2006, Report Prepared for NOAA/NMFS Office of Sustainable Fisheries, NOAA-Michigan Sea Grant Program, FPS (Leland: FPS, 2009 revised edition), 24.

[17] Bob McNabb, "Requiem in Leland," *Grand Rapids Press*, 26 January 1975; Jim VerSnyder used the phrase "unfair deal" in Holmes and Stewart, 31; Tom Dammann, "Big Lake Fishing Permit Lottery 'More a Funeral.'" Newspaper article, Bill Carlson Collection, FPS [April 1975?].

[18] Bill Carlson, interview, 1999.

[19] Marilyn Stallman, interviewed by Amanda Holmes for FPS, 27 November 2007, Lake Leelanau, Michigan.

Fishing Generations

[1] Piet Bennett, "Long Family Tradition Nearing End for Many Lake Michigan Fishermen," *Ludington Daily News*, 27 October 1977.

[2] Buckler, interview, 1991.

[3] Brian Price, interviewed by Daniel Stewart for FPS, 21 May 2008, Leland, Michigan.

[4] Page, *The Traverse Region*, 244-245.

[5] James M. McClurken, *Gah-Baeh-Jhagwah-Buk (The Way It Happened): A Visual Cultural History of the Little Traverse Bay Bands of Odawa* (East Lansing: Michigan State University Museum, 1991), 50; Bogue, *Fishing the Great Lakes*, 6-7; Chiarappa, "Great Lakes Commercial Fishing Architecture," 220.

[6] James McClurken quoted in Cymbre Sommerville Foster, "Fishing for a living,"

Leelanau Enterprise, 22 February 2001; Dick Ristine, interviewed by Amanda Holmes for FPS, 20 August 2008, Leland, Michigan.

[7] Leland Historic District working file; "The Charlevoix Fisheries," *Grand Traverse Herald,* 18 December 1888; Paul G. Connors, "America's Emerald Isle, "The Cultural Invention of the Irish Fishing Community on Beaver Island" (PhD diss., Loyola University, 1999), 311; Buckler, interview, 1991.

[8] Bill Carlson, interviewed by Amanda Holmes for FPS, 13 August 2007, Leland, Michigan.

[9] Ristine, interview, 2008.

[10] John C. Mitchell, *Wood Boats of Leelanau: A Photographic Journey* (Leelanau Historical Society, 2007), 89-93. *Traverse Bay Eagle* account of the *Tiger* and Harting's death also quoted in Mitchell, 92-93.

[11] Ristine, interview, 2008.

[12] Mitchell, *Wood Boats of Leelanau,* 65.

[13] The *Macknac* is listed as Oscar Price's first boat in his obituary. The spelling may be a typographical error for "Mackinaw" boat, or it may be the actual boat name.

[14] *Leelanau Enterprise,* 5 March 1931; "Last Rites Saturday for Warren Price," *Leelanau Enterprise,* 16 August 1956.

[15] Ristine, interview, 2008.

[16] Bruce Price, interview, 2010.

[17] *Leelanau Enterprise,* 10 September 1942; Bruce Price, interview, 2010.

[18] Ristine, 2008; Barbara Gentile, interviewed by Amanda Holmes for FPS, 24 September 2010, Leland, Michigan.

[19] *Leelanau Enterprise,* 14 March 1940; *Leelanau Enterprise,* 13 December 1906, notes that in 1906 Will Buckler moved his family to the mainland at Leland, a date that conflicts with Roy Buckler's oral recollections; Buckler, interview, 1991.

[20] Alan Priest, interviewed by Michael J. Chiarappa, for the Fish for All Project, courtesy of Michael J. Chiarappa and the Great Lakes Research Library, South Haven, Michigan, 29 May 1999, Leland, Michigan; Carlson and Buckler, interview, 1987.

[21] Eric MacDonald with Arnold R. Alanen, "*Tending a Comfortable Wilderness,*" *A History of Agricultural Landscapes on North Manitou Island, Sleeping Bear Dunes National Lakeshore, Michigan* (Omaha, NB: Midwest Field Office, U.S. Department of the Interior, National Park Service, 2000), 160; *Leelanau Enterprise,* 28 July 1921.

[22] Leland Harbor, 1935, 13; *Leelanau Enterprise,* 12 February 1931.

[23] Bruce Price, interview, 2010.

[24] Herbert Nelson, interviewed by Laurie Kay Sommers for FPS, 26 September 2010, Leland, Michigan.

[25] Dick Carlson, interviewed by Laurie Kay Sommers for FPS, 30 November 2010, Grand Rapids, Michigan.

[26] Richard A. Neumann, architect, *Steffens & Stallman Shanty Preservation Plan,* with historical information by Amanda Holmes (Petoskey, Michigan, 2009); "Leland Fishermen

Appear in Movies at Traverse City," *Leelanau Enterprise*, 30 September 1926.

[27] Marilyn Stallman, interview, 2007; Weist, "'Fisherman 'Calls It a Day'"; Bill Carlson, interview, 9 August 2007.

[28] David Grath, interviewed by Amanda Holmes for FPS, 28 December 2007, Northport, Michigan.

[29] Bill Carlson, interview, 13 August 2007; Ann Steffens, phone conversation with Laurie Kay Sommers, 14 September 2010. The limited entry rule, enforced retroactively, forced part-time commercial fishermen out of business by only renewing licenses for those fishermen who logged a certain number of hours and invested at least a designated amount of money in commercial fishing. The rule was designed to protect the fish population and reduce conflict between sport and commercial fishermen. The legislature received authority to manage the fishery on this principle in 1967; the law was imposed in 1969.

[30] "Carlson Passes Away at Leland Home, at age of 81," *Leelanau Enterprise*, 16 September 1935; Rita Hadro Rusco, *North Manitou Island, Between Sunrise and Sunset* (Self-Published, 1991), 54-55; Mark Carlson, interviewed by Laurie Kay Sommers for FPS, 24 August 2010, Leland, Michigan.

[31] "Death Claims Old Fisherman," *Leelanau Enterprise*, 23 June 1949.

[32] Carlson and Buckler, interview, 1987.

[33] Bill Carlson, interview, 1999.

[34] Carlson and Buckler, interview, 1987.

[35] Ibid.

[36] Rita Carlson, interviewed by Laura Quackenbush for Leelanau Historical Society, 1993, Leland, Michigan.

[37] Carlson and Buckler, interview, 1987.

[38] Rita Carlson, interview, 1993.

[39] Ibid.

Fish Boats of Wood and Steel

[1] Brian Price, interview, 2008.

[2] Grath, interview, 2007.

[3] Timothy Cochrane and Hawk Tolson, *A Good Boat Speaks for Itself, Isle Royale Fishermen and Their Boats* (Minneapolis and London: University of Minnesota Press, 2002), xi.

[4] Buckler, interview, 1991.

[5] Carlson and Buckler, interview, 1987.

[6] Cochrane and Tolson, *A Good Boat Speaks for Itself*, xi.

[7] Information from Telgard's obituary, *Northport Leader* 30 June 1934; Mitchell, *Wood Boats of Leelanau*, 120. The list of Telgard's boats is compiled from early *Leelanau Enterprise* references and Mitchell's *Wood Boats of Leelanau*.

[8] Buckler, interview, 1991.

[9] "New Fish boat Launched in Harbor Here, *Nu Deal* joins fleet, Mort Decker does honors as tug slips into water," *Leelanau Enterprise*, 13 September 1934; Bruce Price, interview, 2010.

[10] Jed Jaworsi, "*Diamond* Wreck brings Good Will out of caring townsfolk," *Glen Arbor Sun*, 28 July 2005.

[11] Bogue, *Fishing the Great Lakes*, 258; Michael J. Chiarappa and Kristin Szylvian, *Fish For All, An Oral History of Multiple Claims and Divided Sentiment on Lake Michigan* (East Lansing: Michigan State University Press, 2003), 110, 114.

[12] *Leelanau Enterprise*, (27 February [1958?], Janice Sue Kiessel Collection, FPS.

[13] "Two New Great Lakes Fish Tugs," *Fishing Gazette*, April 1959, 94D-94E. A "skeg" is a tapering or projecting stern section of a vessel's keel, which protects the propeller and supports the rudder.

[14] Brian Price, interview, 2008.

[15] "Two New Great Lakes Fish Tugs," 94D-94E; Glenn Garthe, interviewed by Amanda Holmes for FPS, 15 November 2007, Leland, Michigan.

[16] Garthe, interview, 2007.

[17] Al Parker, "Leland Loves Iconic Fishing Tug," *Traverse City Record Eagle*, 7 September 2008.

[18] Brian Price, interview, 2008; Dammann, "Big Lake Fishing Permit Lottery 'More a Funeral.'"

[19] Dammann, "Big Lake Fishing Permit Lottery 'More a Funeral.'"

[20] Joy Lang, interviewed by Michael Chiarappa for Fish for All Project, courtesy of Michael J. Chiarappa and the Great Lakes Research Library, South Haven, 29 May 1999, Leland, Michigan; Nicholas Lederle, interviewed by Daniel Stewart for FPS, 14 December 2009, Leland, Michigan; Al Campbell, "First at Leland Harbor in 49 years, New hand-made fishing boat is launched," *Leelanau Enterprise*, 1 April 1982.

[21] Ed Peplinski, interviewed by Teresa Scollon for FPS, 28 September 2008, Centerville Township, Michigan; Priest, interview, 2009.

"Special Link to the Mainland": Fishtown's Manitou Ferry and Mail Boat

[1] *Leelanau Enterprise*, 13 February 1930.

[2] Karl W. Detzer "Defies Death to Get Mail Through, Capt. Grosvenor Battles His Way To Ice-Bound Isle Twice a Week," *Detroit News*, 8 February 1931; Mike Grosvenor, interviewed by Daniel Stewart for FPS, 9 October 2008, Leland, Michigan.

[3] MacDonald and Alanen, "*Tending a Comfortable Wilderness*," 381-387.

[4] Robert H. Ruchhoft, *Exploring North Manitou, South Manitou High and Garden Islands of the Lake Michigan Archipelago* (Cincinnati, OH: self published, 1991). Ruchhoft asserts that in the late nineteenth century, "The second largest landowner then was an island resident named Gottlieb Patek, whose 4,000 acres represent another 38 percent of the total area of the island. Patek was one of the earliest to establish and operate a ferry and mail boat to the mainland. He sold his land to a large lumbering firm named Smith and Hull in the beginning of the twentieth century" (p. 184); Rusco, *North Manitou Island, Between Sunrise*

and Sunset, 77; Mitchell, *Wood Boats of Leelanau*, 120.

[5] The *Michigan State Gazetteer* first lists Tracy Grosvenor as captain in the 1921-1922 edition. The Grosvenor family lists the date "1917" on the sign for Manitou Transit; however, in the *Leelanau Enterprise*, the date is given as 1915, based on an interview with George Firestone Grosvenor, son of Tracy (see Tom Montgomery, "Manitou ferries 'carryin' the mail' for 71 years," *Leelanau Enterprise* Looking Back Issue, 27 February 1986, section 3, p. 3). Mike Grosvenor, in his 9 October 2008 interview with Daniel Stewart, gives the date of 1913.

[6] MacDonald and Alanen, "*Tending A Comfortable Wilderness*," 51-55, 403.

[7] Montgomery, "Manitou ferries"; Detzer "Defies Death to Get Mail Through."

[8] David Peterson, *Erhardt Peters "Loving Leland" A Pictorial History* (Ludington: Black Creek Press, 2004), 31; Detzer, "Defies Death to Get Mail Through."

[9] Detzer, "Defies Death to Get Mail Through."

[10] Grosvenor, interview, 2008.

[11] Ibid.

[12] Detzer, "Defies Death to Get Mail Through."

[13] "Postman Braves the Lake Alone," *Leelanau Enterprise*, [1958?], Mike and Beth Grosvenor Collection, FPS, Leland, Michigan.

[14] Boats since the *Smiling Thru* include the following: the fish tug *Bonnie Lass* (John Maleski's fish tug—probably purchased from the Maleski estate in 1950); the *Mollie B* (a fifty-six foot World War II era landing craft used from about 1970-1975 and bought from Northwest Michigan College to transport freight off the Manitous after the creation of Sleeping Bear Dunes National Lakeshore); the *Megan*, the *Manitou Isle*, the *Marineland* (a Munising boat leased for a year about 1967 and returned as not suitable for rough water); the *Island Clipper* (used from about 1967 through 1980 for charter fishing and passengers); and the *Mishe-Mokwa* (the largest boat in the history of the ferry, purchased in 1981).

[15] Detzer, "Defies Death to Get Mail Through."

[16] Grosvenor, interview, 2008.

[17] Ibid.

Brotherhood of the Lake

[1] VerSnyder, interview, 2008.

[2] Carlson and Buckler, interview, 1987.

[3] VerSnyder, interview, 2008.

[4] Clay Carlson, interviewed by Amanda Holmes for FPS, 6 December 2007, Leland, Michigan.

[5] VerSnyder, interview, 2008.

[6] Garthe, interview, 2007.

[7] Bill Carlson, interview, 1999; VerSnyder interview, 2008.

[8] VerSnyder, interview, 2010.

[9] Clay Carlson, interview, 2007.

[10] VerSnyder, interview, 2008.

[11] Chiarappa, "Great Lakes Commercial Fishing Architecture," 218, 224.

[12] Brian Price, interview, 2008; VerSnyder interview, 2008.

[13] VerSnyder, interview, 2008.

[14] Brian Price, interviewed by Clair Gornowicz with Abraham Hohnke for Fish for All Project, courtesy of Michael J. Chiarappa and the Great Lakes Research Library, South Haven, 28 May 1999.

[15] Buckler, interview, 1991.

[16] Ibid.

[17] Dick Carlson, interview, 2010.

[18] VerSnyder, interview, 2008.

[19] Clay Carlson, interview, 2007.

[20] VerSnyder, interview, 2010.

[21] Ibid.

[22] Ibid.

[23] Ibid.

[24] Dick Carlson, interview, 2010.

[25] VerSnyder, interview, 2010.

[26] "Norman Price Takes Cold Swim in Lake," *Leelanau Enterprise*, 20 December 1928.

[27] Carlson and Buckler, interview, 1987.

[28] VerSnyder, interview, 2010.

[29] "Fish Boat Lonely as a Cloud," *Leelanau Enterprise*, 23 March 1950.

Shanties and Shore Work

[1] Bill Carlson, interview, 9 August 2007.

[2] Garthe, interview, 2007.

[3] Claudia Goudschaal, interviewed by Amanda Holmes for FPS, 12 December 2007, Leland, Michigan; Bill Carlson interview, 9 August 2007.

[4] Garthe, interview, 2007.

[5] Glenn Garthe, interviewed by Laurie Kay Sommers for FPS, 27 July 2010, Leland, Michigan.

[6] Bill Carlson, interview, 1999.

[7] Garthe, interview, 2007.

[8] Ibid.

[9] Bill Carlson, interview, 9 August 2007; Garthe, interview, 2007.

[10] Nelson, interview, 2010.

[11] Dick Carlson, interview, 2010.

[12] *Leelanau Enterprise*, 8 July 1926.

[13] *Grand Rapids Herald*, 20 July 1958.

[14] Grosvenor, interview, 2008.

[15] Garthe, interview, 2007.

[16] Garthe, interview, 2007.

[17] Ibid.

[18] Charlie Hall, interviewed by Laurie Kay Sommers for FPS, 28 June 2010, Leland, Michigan.

Ice Houses and the Community Ice Harvest

[1] "Leland Puts Up Ice For Summer, Annual Harvest Begun Yesterday, Later Than Usual," *Leelanau Enterprise*, 19 February 1931.

[2] "Lake Leelanau Frozen Night of January 31," *Leelanau Enterprise*, 4 February 1931.

[3] "Leland Puts Up Ice For Summer."

[4] Oswald Cordes, interviewed by Mary Ellen Hadjisky for Leelanau Historical Society, 14 January 1992, Leland, Michigan.

[5] "Leland Puts Up Ice For Summer"; Eloise Telgard Fahs, interviewed by Teresa Scollon for FPS, 7 October 2009, Leland, Michigan; Cordes, interview, 1992.

[6] "Leland Puts Up Ice For Summer."

[7] Mike Heuer, "Ice was 'harvested' before electricity," *Leelanau Enterprise*, 27 February 1999.

[8] Cordes, interview, 1992; Bill Carlson and Jim VerSnyder, interviewed by Laurie Sommers for FPS, 25 May 2010, Leland, Michigan.

[9] Bill Carlson, interview, 9 August 2007; Garthe interview, 2007.

[10] Bill Carlson, interview, 9 August 2007.

[11] Roy Buckler, quoted in Heuer, "Ice was 'harvested' before electricity."

Stories and Shenanigans

[1] *Leelanau Enterprise*, 29 September 1904

[2] Grosvenor, interview, 2008.

[3] Priest, interview, 1999.

[4] John Westol, interviewed by Amanda Holmes for FPS, 29 July 2008, Leland, Michigan.

[5] Brian Price, interview, 2008.

[6] Grosvenor, interview, 2008.

[7] Ibid.

[8] VerSnyder, interview, 2010; Lang interview, 1999.

[9] Priest, interview, 1999.

[10] VerSnyder, interview, 2008.

[11] Ibid.

Saving Fishtown

[1] Bill Carlson, interview, 9 August 2007.

[2] Goudschaal, interview, 2007.

[3] Mitchell, *Wood Boats of Leelanau*, 88; *Leelanau Enterprise*, 1 July 1897, 15 July 1897.

[4] Littell, *Leland, An Historical Sketch*, 34; Chandler Olds Higley, "Recollections of Leland As a Youth and a Young Man," David P. Higley, ed., (Leland: Published by David P. Higley, June, 2010), 1.

[5] *Leelanau Enterprise*, Special History Edition, 14 December 1944; *Leelanau Enterprise*, 23 May 1907.

[6] Caroline Brady, interviewed by Laurie Kay Sommers for FPS, 31 July 2010; Higley, "Recollections of Leland," 14.

[7] Higley, "Recollections of Leland," 8; Elizabeth Wiese, interviewed by Laurie Kay Sommers for FPS, 28 July 2010, Leland, Michigan; Higley, "Recollections of Leland," 8.

[8] Barbara Gentile, interviewed by Amanda Holmes for FPS, 24 September 2010, Leland, Michigan.

[9] Ristine, interview, 2008.

[10] Goudschaal, interview, 2007.

[11] Jane Shermerhorn, "Summer Colony in Leland Finds Life is Easy and Gay," *Detroit News*, 21 July 1957.

[12] Hall, interview, 2010.

[13] Barbara Goodbody, phone conversation with Laurie Kay Sommers, 12 August 2010.

[14] Dan C. Appel, interviewed by Laurie Kay Sommers for FPS, 30 July 2010, Leland, Michigan; Ken Krantz, interviewed by Laurie Kay Sommers for FPS, 5 February 2001, Suttons Bay, Michigan.

[15] Lawrence M. Sommers et al., *Fish in Lake Michigan, Distribution of Selected Species*, Michigan Sea Grant Advisory Program (East Lansing: Michigan State University Publications, 1981), 16; Jim Munoz, interviewed by Amanda Holmes for FPS, 15 January 2008, Leland, Michigan.

[16] Jack Duffy, interviewed by Amanda Holmes for FPS, 2 January 2008, Leland, Michigan.

[17] Bill Carlson, interview, 13 August 2007.

[18] Bill Carlson, interview, 9 August 2007.

[19] Bill Carlson, interviewed by Amanda Holmes, 9 August 2007.

[20] Mark Carlson, interview, 2010; Bill Carlson, interview, 13 August 2007; Bill Carlson, quoted in Bill O'Brien, "Historic Fishtown faces sale, uncertain future," *Traverse City Record Eagle*, 12 June 2005.

[21] Alan Priest, quoted in *FPS Newsletter* 3, no. 2 (Fall 2009).

Selected Bibliography

Fish for All Interviews (Michael J. Chiarappa *et al.*)

Anderson, Scott. Interviewed by Michael J. Chiarappa, Fish for All Project, 26 May 1999, courtesy Michael J. Chiarappa and the Great Lakes Research Library, South Haven.

Carlson, William (Bill). Interviewed by Michael J. Chiarappa, Fish for All Project, 26-27 May 1999, Leland, Michigan, courtesy Michael J. Chiarappa and the Great Lakes Research Library, South Haven.

Lang, Joy. Interviewed by Michael J. Chiarappa, Fish for All Project, 29 May 1999, courtesy Michael J. Chiarappa and the Great Lakes Research Library, South Haven.

Price, Brian. Interviewed by Clair Gornowicz with Abraham Hohnke, Fish for All Project, 28 May 1999, courtesy Michael J. Chiarappa and the Great Lakes Research Library, South Haven.

Priest, Alan. Interviewed by Michael J. Chiarappa, Fish for All Project, 29 May 1999, Cedar, Michigan, courtesy Michael J. Chiarappa and the Great Lakes Research Library, South Haven.

Fishtown Preservation Society Interviews

Anderson, Phil. Interviewed by Amanda Holmes, 16 April 2008, Leland, Michigan.

Appel, Daniel (Dan) C.. Interviewed by Laurie Kay Sommers, 30 July 2010, Leland, Michigan.

Brady, Caroline. Interviewed by Laurie Kay Sommers, 31 July 2010, Leland, Michigan.

Carlson, William (Bill). Interviewed by Amanda Holmes, 9 and 13 August 2007, Leland, Michigan.

Carlson, William (Bill) and James VerSnyder. Interviewed by Laurie Kay Sommers, 25 May 2010, Leland, Michigan.

Carlson, Mark. Interviewed by Laurie Kay Sommers, 25 August 2010, Leland, Michigan.

Carlson, Richard (Dick). Interviewed by Laurie Kay Sommers, 30 November 2010, Grand Rapids, Michigan.

Craig, Susann. Interviewed by Amanda Holmes, 3 August 2009, Leland, Michigan.

Derusha, James (Jim). Interviewed by Daniel Stewart, 30 January 2010, Traverse City, Michigan.

Duck, Berkley, III. Interviewed by Daniel Stewart, 19 September 2009, Leland, Michigan.

Duffy, John A. (Jack). Interviewed by Amanda Holmes, 2 January 2008, Leland, Michigan.

Fahs, Eloise Telgard. Interviewed by Teresa Scollon, October 2009, Leland, Michigan.

Garthe, Glenn. Interviewed by Amanda Holmes, 15 November 2007, Leland, Michigan.

Garthe, Glenn. Interviewed by Laurie Kay Sommers, 27 July 2010, Leland, Michigan.

Gentile, Barbara, and Betty English. Interviewed by Amanda Holmes, 24 September 2010, Leland, Michigan.

Goudschaal, Claudia D. Interviewed by Amanda Holmes,12 December 2007.

Grath, David P. Interviewed by Amanda Holmes, 28 December 2007, Northport, Michigan.

Grosvenor, Michael (Mike). Interviewed by Daniel Stewart, 9 October 2008, Leland, Michigan.

Hall, Charles (Charlie). Interviewed by Laurie Kay Sommers, 28 June 2010, Leland, Michigan.

Hall, Michael (Mike). Interviewed by Amanda Holmes, 30 August 2008, Leland, Michigan.

Hummel, Roger A. Interviewed by Laurie Kay Sommers, 18 August 2010, Suttons Bay, Michigan.

Kelly, Tom. Interviewed by Teresa Scollon, 7 October 2008, Suttons Bay, Michigan.

Krantz, Kenneth (Ken). Interviewed by Laurie Kay Sommers, 5 February 2011, Suttons Bay, Michigan.

Lederle, Nicholas (Nick). Interviewed by Daniel Stewart, 14 December 2009, Leland, Michigan.

Lederle, Nicholas and Susann. Interviewed by Laurie Kay Sommers, 28 July 2010, Leland, Michigan.

Munoz, Jim. Interviewed by Amanda Holmes, 15 January 2008, Leland, Michigan.

Nelson, Herbert (Herb). Interviewed by Laurie Kay Sommers, 27 September 2010, Lake Leelanau, Michigan.

Peplinski, Ed. Interviewed by Teresa Scollon, 8 September 2008, Centerville Township, Michigan.

Price, Bruce. Interviewed by Laurie Kay Sommers, 19 August 2010, Lake Leelanau, Michigan.

Priest, Alan. Interviewed by Daniel Stewart, 27 February 2009, Cedar, Michigan.

Price, Brian. Interviewed by Daniel Stewart, 21 May 2008, Leland, Michigan.

Ristine, Dick. Interviewed by Amanda Holmes, 20 August 2008, Leland, Michigan.

Stallman, Leo Jr. Interviewed by Laurie Kay Sommers, 27 August 2010, Traverse City, Michigan.

Stallman, Marilyn. Interviewed by Amanda Holmes, 27 November 2007, Lake Leelanau, Michigan.

VerSnyder, James (Jim). Interviewed by Daniel Stewart, 12 May 2008, Leland, Michigan.

VerSnyder, James (Jim). Interviewed by Laurie Kay Sommers, 26 September 2010, Leland, Michigan.

Westol, John. Interviewed by Amanda Holmes, 29 July 2008, Leland, Michigan.

Wiese, Elizabeth. Interviewed by Laurie Kay Sommers, 27 July 2010, Leland, Michigan.

Leelanau Historical Society Interviews

Buckler, Roy. Interviewed by Laura Quackenbush, 1991, Leland, Michigan.

Buckler, Roy and William Lester (Pete) Carlson, 1987, Leland, Michigan.

Carlson, Rita. Interviewed by Laura Quackenbush, 17 February 1993, Leland, Michigan.

Cordes, Oswald (Ozzie). Interviewed by Mary Ellen Hadjisky, 14 January 1992, Leland, Michigan.

Grosvenor, George. Interviewed by Betty Mann, 27 May 1993, Leland, Michigan.

Selected Publications

Anderson, Charles M. *Isle of View: A History of South Manitou Island*. Frankfort, Michigan: J. B. Publications, 1979.

Brady, Caroline. "Looking for Uncle Jack." In *J. Ottis Adams 1851-1927, American Impressionist in Leland*. Published in conjunction with the exhibition of the same name, shown at the Dennos

Museum Center, Northwest Michigan College, Traverse City, Michigan, 12 September through 4 November 1993.

Bogue, Margaret Beattie. *Fishing the Great Lakes: An Environmental History, 1783-1933*. Madison: University of Wisconsin Press, 2000.

Bowling Green State University. Historical Collections of the Great Lakes, Great Lakes Vessels Online Index. Bowling Green, OH, 5 July 2003, http://ul.bgsu.edu/cgi-bin/xvsl2.cgi.

Cashman, William. "The Rise and Fall of the Fishing Industry." *The Journal of Beaver Island History* 1 (1976): 69-87.

Chiarappa, Michael J. "Great Lakes Commercial Fishing Architecture: The Endurance and Transformation of a Region's Landscape/Waterscape." In *Perspectives in Vernacular Architecture: Building Environments* 10 (2005), edited by Kenneth A. Breisch and Alison K. Hoagland, 217-232.

Chiarappa, Michael J. and Kristin M. Szylvian. "Heeding the Landscape's Usable Past: Public History in the Service of a Working Waterfront." *Perspectives in Vernacular Architecture: Building Environments* 16, no. 2 (2009): 86-113.

—. *Fish for All: An Oral History of Multiple Claims and Divided Sentiment on Lake Michigan*. East Lansing: Michigan State University Press, 2003.

Closser Associates, Real Estate Appraisal. *Appraisal of Fishtown, Leland, Michigan*. February 2005.

Cochrane, Timothy. "Commercial Fishermen and Isle Royale: A Folk Group's Unique Association with Place." In *Michigan Folklife Reader*, edited by C. Kurt Dewhurst and Yvonne Lockwood, 87-105. East Lansing: Michigan State University Press, 1988.

Cochrane, Timothy, and Hawk Tolson. *A Good Boat Speaks for Itself: Isle Royale Fishermen and Their Boats*. Minneapolis: University of Minnesota Press, 2002.

Connors, Paul G. "America's Emerald Isle: The Cultural Invention of the Irish Fishing Community on Beaver Island." PhD diss., Loyola University, 1999.

Dickinson, Frederick W. *A Short History of the Leland Iron Works*. Annotated by Harley W. Rhodehamel. Leland: Leelanau Historical Society, 1996.

Doherty, Robert. *Disputed Waters: Native Americans and the Great Lakes Fishery*. Lexington: University Press of Kentucky, 1990.

Dunbar, Willis S. *Michigan: A Guide to the Wolverine State*. Rev. ed. by George S. May. Grand Rapids: William B. Eerdmans Publishing Company,1980.

Edwards, Elizabeth. "The Fisherman Who Saved Fishtown." *Traverse Magazine* (July 2007). Web update, 4 March 2008. http://www.mynorth.com/My-North/July-2007/The-Fisherman-Who-Saved-Fishtown/.

Ferris, Charles E. *Atlas of Leelanau County*. Knoxville, TN and Solon, MI: S. B. Newman, Printers, J. White, Sole Agent, 1900.

Fishing Gazette. "Two New Great Lakes Fish Tugs" (April 1959), 94D-94E.

Fishtown Preservation Society Newsletters, 2007-2011.

Goodschaal, Claudia D. *Destination: Leelanau, Boats, Sailing Leelanau Waters, 1835-1900*. Omena, MI: Self-published, 2009.

Grand Traverse Area Genealogical Society, Inc. *Cemeteries of Leelanau County, Michigan*. Traverse City, Michigan, 2006.

Hayes, E. L. and Titus, C. O. and Co. *Atlas of Leelanau County, Michigan, compiled and drawn for the publisher by E.L. Hayes*. Philadelphia: C. O. Titus Co., 1881.

Higley, Chandler Olds. *Recollections of Leland As a Youth and a Young Man*. Edited and self-published by David P. Higley. Leland: June 2010.

Holmes, Amanda J. and Daniel W. Stewart. *"We're All Brothers Out There": Voices from the Leland Fishery at Fishtown, 1940-2006,* rev. ed. Report Prepared for NOAA/NMFS Office of Sustainable Fisheries, NOAA-Michigan Sea Grant Program, and Fishtown Preservation Society. Leland: Fishtown Preservation Society, 2009.

Hubbard, Bela. Papers. Peninsula Coast Survey, Detroit to Chicago. Box 1, Survey Notebook 2, Field Notebooks May 19-July 24, 1838. Bentley Historical Library, University of Michigan.

JJR, LLC and HopkinsBurns Design Studio. *Fishtown Site Study, Design and Master Plan.* Ann Arbor: January 2009.

Kelly, Thomas M. *The Grand Traverse Bay Fishery in 1986.* Prepared for the Grand Traverse Band of Ottawa and Chippewa Indians, November 1987.

Lafferty, William. "Anatomy of a Fish Tug: the Jean R." *Lake Michigan Maritime Marginalia* 1, no. 3 (1999). http://www.wright.edu/~william.lafferty/vln3intro.htm.

Leelanau Community Cultural Center and Michigan State University Department of Art and Art History. *MSU Artists Celebrating Past and Present, 1939-1989, The Old Art Building in Leland.* Published in conjunction with an exhibition of the same name, shown in the Old Art Building, Leland, Michigan, 2007.

Littell, Edmund M. *100 Years in Leelanau.* Leland: Leelanau County Prospectors Club, 1965.

Littell, Joseph. *Leland, an Historical Sketch.* Indianapolis Printing Company, 1920.

McClurken, James M. Affidavit. United States District Court for the Western District of Michigan, Grand Traverse Band of Ottawa and Chippewa Indians, Plaintiff, File No. 1-94-CV-707, Hon. Richard A. Enslen, 14 April 1995.

—. *Gah-Baeh-Jhagwah-Buk (The Way It Happened): A Visual Cultural History of the Little Traverse Bay Bands of Odawa.* East Lansing: Michigan State University Museum, 1991.

—. "We Wish to be Civilized: Ottawa-American Political Contests on the Michigan Frontier." PhD diss., Michigan State University, 1988.

MacDonald, Eric with Arnold R. Alanen. *Tending a "Comfortable Wilderness," A History of Agricultural Landscapes on North Manitou Island, Sleeping Bear Dunes National Lakeshore, Michigan.* Omaha, NB: Midwest Field Office, U.S. Department of the Interior, National Park Service, 2000.

Michigan Sea Grant. "Know Your Nets." http://www.miseagrant.umich.edu/nets/.

Michigan State Board of Fish Commissioners. *Fishery Reports of Michigan Fisheries.* Second District, 1897-1908, Archives of Michigan, Lansing, Michigan.

Michigan State Gazetteer and Business Directory. Detroit: R. L. Polk & Co., 1860-1931.

Michigan Yearbook. "Brush Strokes at Leland, Michigan." 1960.

Milner, James W. "Report on the Fisheries of the Great Lakes: The Result of Inquiries Prosecuted in 1871 and 1872." In U.S. Commission of Fish and Fisheries. *Report, Fisheries of the Great Lakes 1872-1873.* Appendix A, 42d Cong., 3d sess., 1872. S Misc. Doc. 74 (Serial 1547).

Mitchell, John C. *Wood Boats of Leelanau: A Photographic Journal.* Leland: Leelanau Historical Society, 2007.

Moore, Alan William. "Michigan Historic Fisheries: Policy and Preservation Considerations Involved in Their Establishment in a Michigan Shoreline Community." M.S. thesis, University of Michigan, 1975.

National Register of Historic Places. "Leland Historic District." Department of the Interior, National Park Service, 20 November 1975.

Neumann, Richard A., architect. *Steffens & Stallman Shanty Preservation Plan.* With historical information by Amanda Holmes. Petoskey, MI: 2009.

Otwell Mawby, P.C., Consulting Engineers. *Historic Fishtown/Carlson Properties Baseline Environmental Assessment.* Traverse City, MI: 2007.

—. *Manitou Transit Company Property, Lot 7, West River Street, Village of Leland, Leelanau County, Phase I Environmental Site Assessment.* Traverse City, MI: March 2010.

Page, H. R. & Co. The *Traverse Region, Historical and Descriptive, and Portraits and Biographical Sketches.* Chicago: H. R. Page & Co., 1884.

Pelizzari, Patty and Jacqueline Shinners. "Introduction." In *J. Ottis Adams 1851-1927, American Impressionist in Leland.* Published in conjunction with the exhibition of the same name, shown at the Dennos Museum Center, Northwest Michigan College, Traverse City, Michigan, 12 September through 4 November 1993.

Petersen, David. *Erhardt Peters, Loving Leland, Michigan.* Ludington: Black Creek Press, 2004.

Prothero, Frank. *Men 'n Boats: The Fisheries of the Great Lakes.* Port Stanley, ON: Great Lakes Fisherman, 1975.

Price, Brian, and Thomas M. Kelley. *Fishes of the Grand Traverse Bay Region.* Ann Arbor: Michigan Sea Grant Program, 1976.

Rastetter, Bill. 1836 Treaty — Time Line re: Reserved Usufruct Rights. Prepared for Grand Traverse Band members. http://turtletalk.files.wordpress.com/2007/09/1836-treaty-timeline-obh-current-version.pdf.

Ruchhoft, Robert H. *Exploring North Manitou, South Manitou High and Garden Islands of the Lake Michigan Archipelago.* Cincinnati, OH: Self-published, 1991.

Rusco, Rita Hadra. *North Manitou Island, Between Sunrise and Sunset.* Self-published, 1991.

Smith, Hugh M. and Merwin-Marie Snell. "Review of the Fisheries of the Great Lakes in 1885." In U.S. Commission of Fish and Fisheries, *Report*, 1887. Appendix. 50th Cong., 2d sess., 1889. H. Misc. Doc. 133 (Serial 2661).

Sommers, Laurie Kay, Eugene C. Hopkins, Evan Hall, Mark Johnson, and Jessica Neafsey. *The River Runs Through It, Report on Historic Structures and Site Design in the Fishtown Cultural Landscape.* Leland: Fishtown Preservation Society, 2011.

Sommers, Lawrence M., ed. *Fish in Lake Michigan, Distribution of Selected Species.* East Lansing: Michigan State University Publications and Michigan Sea Grant Advisory Program, 1981.

Sprague, Elvin L. and Mrs. George N. Smith, eds. *Sprague's History of Grand Traverse and Leelanau Counties, Michigan.* B. F. Bowen, Publisher: 1903.

Szylvian, Kristin M. "Transforming Lake Michigan into the 'Worlds greatest fishing hole': The Environmental Politics of Michigan's Great Lakes Sport Fishing, 1965-1985." *Environmental History* (2004): 102-27.

Tanner, Helen Hornbeck, ed. *Atlas of Great Lakes Indian History.* Norman: University of Oklahoma Press, 1987.

U.S. Army. "Leland Harbor, Michigan." Letter from the Chief of Engineers, Report of the Board of Engineers for Rivers and Harbors on Review of Reports Heretofore Submitted on Leland Harbor, Mich. With Illustrations, 17 May 1935.

U.S. Army Corps of Engineers. "Survey Report on Leland Harbor, Michigan." U.S. Army Engineer District. Detroit: June 1961.

U.S. Congress. House. "Leland and Empire Harbors, Michigan." Letter from the Acting Secretary of War, 61st Congr., 2d sess, Document 831, 31 March 1910.

U.S. Department of Transportation, United States Coast Guard. *Merchant Vessels of the United States,* 1912, 1920, 1927, 1929, 1937, 1978.

Vrana, Kenneth, ed. *Inventory of Maritime and Recreation Resources of the Manitou Passage Underwater Preserve.* East Lansing: Center for Maritime and Underwater Resource

Management, Department of Parks, Recreation and Tourism Resources, Michigan State University, 1995.

Wadsworth, Abram S. *Map of Leelanau County* by Abram S. Wadsworth, Deputy Surveyor of an Area Geological Survey, 1851.

Wakefield, Lawrence, ed. *A History of Leelanau Township*. Northport, MI: Friends of the Leelanau Township Library, 1983.

Wakefield, Lawrence and Lucille Wakefield. *Sail & Rail: A Narrative History of Transportation in the Traverse City Region*. Traverse City: Village Press, 1980.

Weeks, George. *Mem-ka-weh, Dawning of the Grand Traverse Band of Ottawa and Chippewa Indians*. Grand Traverse Band of Ottawa and Chippewa Indians, 1992.

White, Richard. "Ethnohistorical Report on the Grand Traverse Ottawas." History Department, Michigan State University, n.d.

Winchell, Alexander. *A Report on the Geological and Industrial Resources of the Counties of Antrim, Grand Traverse, Benzie and Leelanaw in the Lower Peninsula of Michigan*. Ann Arbor: Dr. Chase's Steam Printing House, 1866.

Works Progress Administration. *Michigan: A Guide to the Wolverine State*. Oxford University Press, 1941.

Selected Newspapers

Bennett, Piet. "Long Family Tradition Nearing End for Many Lake Michigan Fishermen." *Ludington Daily News*, 27 October 1977.

Christian Science Monitor. "Commercial, Sport Anglers Feud Over Dwindling Fish." 6 September 1994.

—. "A Great Lakes Fisherman Learns Ancestral Lesson." 6 September 1994.

Campbell, Al. "First at Leland Harbor in 49 years, New hand-made fishing boat is launched," *Leelanau Enterprise*, 1 April 1982.

Dammann, Tom. "Big Lake Fishing Permit Lottery 'More a Funeral.'" Newspaper article, Bill Carlson Collection, Fishtown Preservation Society [April 1975?].

Detroit News. "Leland's Artistic Tradition Saved, Picturesque Sights Again to Greet Summer Tourists." 9 January 1959.

Detzer, Karl W. "Happy 100th Birthday Leelanau." *Leelanau Enterprise* Centennial Edition, 25 May 1963.

—. "Defies Death to Get the Mail Through, Capt. Grosvenor Battles His Way To Ice-Bound Isle Twice a Week." *Detroit News*, 8 February 1931.

Haven, Kenn. "Harbor Won't Interfere, Leland's Fishtown is Saved." *Detroit Free Press*, 27 July 1966.

Heur, Mike. "Ice was 'harvested' before electricity." *Leelanau Enterprise*, 27 February 1997.

Leelanau County Times "Fishing Trout in Big Lake." 13 November 1947.

Leelanau Enterprise. "Local Trout Fishing Continues Merely Fair." 13 November 1980.

—. "Mackinaw Off Leland, Michigan." October 1960.

—. "Ice Forms Giant Trap for Leland Fish Tugs." 27 February 1958.

—. "Shifting Ice Traps Boats." 25 February 1958.

—. "Steel Tugs in Leland Fleet." 28 October 1958.

—. "Fisherman Calls it a Day." [1954?], Janice Sue Kiessel Collection, Fishtown Preservation Society, Leland, Michigan.

—. "Fish Boat Lonely as a Cloud." 23 March 1950.

—. Special History Edition. 14 December 1944.

—. "Captain Oscar Price, Obituary." 10 September 1942.

—. "Grievances of Fishermen To Be Aired at Meeting." 25 January 1940.

—. "New Fish Boat Launched in Harbor Here, *Nu Deal* joins fleet, Mort Decker does honors as tug slips into water." 13 September 1934.

—. "Commercial Fishermen Turn Attention to Nets." 3 March 1932.

—. "Ice Harvest is Finished After Many Delays." 3 March 1932.

—. "Fishermen Set Their Nets for the Trout Run." 12 November 1931.

—. "Trawling of Trout Spawn Will Commence Soon." 15 October 1931.

—. "Warren Price to Fish Again, Goes into Partnership with William Carlson." 5 March 1931.

—. "Fishermen Set Nets This Week, Oscar Price, Kaapke & Firestone, Went Out Monday." 26 February 1931.

—. "Leland Puts Up Ice for Summer, Annual Harvest Begun Yesterday, Later Than Usual." 19 February 1931.

—. "Three Drift on Lake Michigan for Ten Hours, Henry Steffens Brings in Kaapke, Firestone and Egeler Early Sunday." 12 February 1931.

—. "Fishermen Lift Trout Tuesday, Catch Is Not Large; Stormy Weather Prevails Again." 6 November 1930.

—. "Fishermen Mend Nets This Week to Prepare for Next." 30 October 1930.

—. "Rearing Pond Ready for Fish." 19 June 1930.

—. "Michigan Fishermen Use Over a Thousand Boats." 10 April 1930.

—. "Michigan Leads on Great Lakes Fishing Industry." 27 March 1930.

—. "Leelanau Fishermen Are Receiving Just Praise." 13 February 1930.

—. "Lake Leelanau Frozen Night of January 31, Ice is 4 Inches Thick." 4 February 1931.

—. "Ice Harvest is on This Week." 9 February 1928.

—. "Norman Price Takes Cold Swim in Lake." 20 December 1928.

Lewis, Corrine. "Leland Man Admits There's 'Something Fishy' About His Business." *Preview Community Weekly*, Traverse City (MI), 19 October 1981.

McNabb, Bob. "Requiem in Leland." *Grand Rapids Press*, 26 January 1975.

Montgomery, Tom. "Manitou ferrys 'carrying' the mail' for 71 years." *Leelanau Enterprise* Looking Back Issue, 27 February 1986, sec. 3, p. 3.

O'Brien, Bill. "Historic Fishtown faces sale, uncertain future." *Traverse City Record Eagle*, 12 June 2005.

Olson, Chris. "A Shipshape Goal for Two Historic Tugs." *Leelanau Enterprise*, 30 December 2008.

Parker, Al. "Leland Loves Iconic Fishing Boat." *Traverse City Record Eagle*, 7 September 2008. http://recordeagle.com/local/x75059660/Leland-loves-iconic-fishing-boat/print.

Russell, Mary. "Commercial Fishing Was Once A 'Major Industry' For Some Communities." *Preview Community Weekly*, Traverse City (MI), 21 May 1990.

Shermerhorn, Jane. "Summer Colony in Leland Finds Life is Easy and Gay." *Detroit News*, 21 July 1957.

Sommerville Foster, Cymbre. "Fishing for a living." *Leelanau Enterprise*, 22 February 2001.

Weist, Jan. "A Model Leland Fisherman Casts Lot with Tradition." *Detroit News*, 24 October 1977.

Index

About the Author

Laurie Kay Sommers is a freelance folklorist and historic preservation consultant based in Okemos, Michigan. She has worked on community documentation projects in Michigan, California, Indiana, Georgia, and Florida for organizations such as the Michigan State University Museum and the South Georgia Folklife Project at Valdosta State University. Throughout the past three decades she has taught university courses in folklore and ethnomusicology and developed numerous public arts and humanities programs, including folk festivals, museum exhibits, and documentary radio. Most recently, she worked as the oral historian and folklorist for the Fishtown Preservation Society in Leland, Michigan, as part of the interdisciplinary team that prepared *The River Runs Through It, Report on Historic Structures and Site Design in the Fishtown Cultural Landscape* (2011). *Fishtown: Leland, Michigan's Historic Fishery* grew out of research for this report.

Laurie is the author of various other publications involving history and folklore, among them *Fiesta, Fe, y Cultura, Celebration of Faith, Culture and Community in Detroit's Colonia Mexicana* and *Beaver Island House Party.*